Java数据科学指南

Java Data Science
Cookbook

[加]鲁什迪·夏姆斯（Rushdi Shams）著
武传海 译

人民邮电出版社
北京

图书在版编目（ＣＩＰ）数据

　Java数据科学指南 ／（加）鲁什迪·夏姆斯（Rushdi Shams）著；武传海译. -- 北京：人民邮电出版社，2018.6
　ISBN 978-7-115-48163-4

　Ⅰ．①J… Ⅱ．①鲁… ②武… Ⅲ．①JAVA语言－程序设计－指南 Ⅳ．①TP312.8-62

　中国版本图书馆CIP数据核字(2018)第057957号

版权声明

Copyright ©2017 Packt Publishing. First published in the English language under the title *Java Data Science Cookbook*.
All rights reserved.

本书由英国 Packt Publishing 公司授权人民邮电出版社出版。未经出版者书面许可，对本书的任何部分不得以任何方式或任何手段复制和传播。

版权所有，侵权必究。

- ◆ 著　　　[加] 鲁什迪·夏姆斯（Rushdi Shams）
  　译　　　武传海
  　责任编辑　胡俊英
  　责任印制　沈 蓉　焦志炜
- ◆ 人民邮电出版社出版发行　北京市丰台区成寿路11号
  邮编　100164　电子邮件　315@ptpress.com.cn
  网址　http://www.ptpress.com.cn
  固安县铭成印刷有限公司印刷

- ◆ 开本：800×1000　1/16
  印张：20
  字数：396千字　　　　　2018年 6 月第 1 版
  印数：1－2 400 册　　　2018年 6 月河北第 1 次印刷

著作权合同登记号　图字：01-2017-4028 号

定价：79.00 元
读者服务热线：(010)81055410　印装质量热线：(010)81055316
反盗版热线：(010)81055315
广告经营许可证：京东工商广登字 20170147 号

内容提要

现如今,数据科学已经成为一个热门的技术领域,它涵盖了人工智能的各个方面,例如数据处理、信息检索、机器学习、自然语言处理、数据可视化等。而 Java 作为一门经典的编程语言,在数据科学领域也有着卓越的表现。

本书旨在通过 Java 编程来引导读者更好地完成数据科学任务。本书通过 9 章内容,详细地介绍了数据获取与清洗、索引的建立和检索数据、统计分析、数据学习、信息的提取、大数据处理、深度学习、数据可视化等重要主题。

本书适合想通过 Java 解决数据科学问题的读者,也适合数据科学领域的专业人士以及普通 Java 开发者阅读。

作者简介

Rushdi Shams 毕业于加拿大韦仕敦大学，获得了机器学习应用博士学位，主攻方向是自然语言处理（Natural Language Processing，NLP）。在成为机器学习与 NLP 领域的专家之前，他讲授本科生与研究生课程。在 YouTube 上，他一直运营着一个名为"跟 Rushdi 一起学"（Learn with Rushdi）的频道，并且做得有声有色，该频道主要面向想学习计算机技术的朋友。

感谢我的家人、朋友与同事，谢谢你们不断地给予我支持、鼓励，以及中肯的批评与建议。

此外，还要感谢 Packt 公司的 Ajith 与 Cheryl，谢谢他们自发地与我保持持续的合作关系。

审稿人简介

Prashant Verma 从 2011 年就开始了他的 IT 生涯，当时他是爱立信公司的一名 Java 开发人员，面向的是电信领域。在有了几年的 Java EE 开发经验之后，他转战大数据领域，几乎用过所有流行的大数据技术，比如 Hadoop、Spark、Kafka、Flume、Mongo、Cassandra 等，而且还熟悉 Scala 与 Python 编程语言。目前，他供职于 QA Infotech 公司，是一名首席数据工程师，致力于使用数据分析与机器学习解决 E-Learning 领域中的问题。

此外，Prashant 也是 Packt 出版的 *Apache Spark for Java Developers* 一书的技术审稿人。

首先感谢 Packt Publishing 给我审阅本书的机会，还要感谢我的雇主、家人，谢谢他们在我忙于审读本书时所表现出的耐心。

谨以此书献给我漂亮的妻子 Mah-Zereen 与可爱的女儿 Ruayda！

前言

当今，数据科学是一个专业化的热门领域，它涵盖人工智能的各个方面，比如数据处理、信息检索、机器学习、自然语言处理、大数据、深度神经网络、数据可视化。在本书中，我们将讲解数据科学领域中既流行又智能的技术，这些技术分散在全书各个章节中，涉及70多个问题。

请记住，目前各个领域对高级数据科学家的需求是非常旺盛的，我们主要使用Java编写了本书各个章节，包括那些使用Java编写的著名的、经典的、最新的数据科学库。首先我们介绍数据采集与清洗流程，而后了解一下如何对所获取的数据建立索引以及进行检索。随后，我们讲解数据的统计描述、统计推断及其应用。接着，安排两个连续的章节讲解面向数据的机器学习应用，这些内容是创建智能系统的基础。除此之外，所讲解的内容还包括现代信息检索与自然语言处理技术。大数据是一个新兴的热门领域，本书内容会涉及其中几个方面。并且，我们还会讲解使用深度神经网络进行深度学习的基础知识。最后，我们学习如何使用有意义的视觉方式或图形表示数据以及从数据中获取的信息。

本书面向的读者是那些对数据科学感兴趣，或者打算应用Java数据科学技术来进一步理解底层数据的朋友。

本书内容安排

第1章 "获取数据与清洗数据"，介绍各种读写数据的方法，以及对数据进行清洗去除其中噪声的方法。本章所涉及的数据文件类型广为人知，比如PDF、ASCII、CSV、TSV、XML、JSON。此外，本章还介绍用来提取Web数据的方法。

第2章 "为数据建立索引与搜索数据"，讲解如何使用Apache Lucene为数据建立索引以实现快速检索。本章介绍的技术是现代搜索技术的基础。

第 3 章 "数据统计分析",讲解应用 Apache Math API 进行数据收集乃至分析统计指标的内容。本章还包含一些高级概念,比如统计显著性检验这一标准的工具,科研人员可以通过它将得到的结果与基准数据进行比较。

第 4 章 "数据学习 I",包含使用 Weka 机器学习库进行分类、聚类、特征选择的内容。

第 5 章 "数据学习 II",本章是前一章的后续内容,讲解使用另一个 Java 库——Java-ML 库进行数据导入导出、分类、特征选择的内容。本章还包含使用斯坦福分类器(Stanford Classifier)与 MOA(Massive Online Access)进行基本分类的内容。

第 6 章 "从文本数据提取信息",介绍对文本数据应用数据科学技术以提取信息的方法。内容涉及 Java 核心应用以及 OpenNLP、Stanford CoreNLP、Mallet、Weka 等著名的机器学习库,学习如何使用它们来完成信息提取与检索任务。

第 7 章 "处理大数据",本章涵盖机器学习大数据平台应用的内容,比如 Apache Mahout、Spark-MLib。

第 8 章 "数据深度学习",包含使用 DL4j 库进行深度学习的基础内容,介绍 word2vec 算法、信念网络与自动编码器。

第 9 章 "数据可视化",讲解如何使用 GRAL 包为数据生成具有吸引力的信息展示图表。这个包功能众多,我们只讲解其中最基础的绘图功能。

阅读本书需要具备的知识

在本书中,我们使用 Java 来解决各种实际的数据科学问题。对于那些想了解如何使用 Java 解决问题的朋友,本书所讲解的内容正是他们所需要的。阅读本书需要你具备最基本的 Java 知识,比如懂得 Java 类、对象、方法、实参与形参、异常、导出 JAR 文件等内容。书中给出的代码都配有相应的讲解、介绍以及提示,这有助于各位读者更好地理解它们。对于书中所解决问题的背后原理,大部分我们都不会进行详细讲解,但必要时我们会提供相应的参考内容,以供感兴趣的读者进一步学习。

本书的目标读者

如果你想了解如何使用 Java 解决现实世界中与数据科学相关的问题,那么本书正是为你准备的。在内容覆盖方面，由于本书内容涵盖数据科学的方方面面，因此对于那些正在从事数据学习相关工作，并寻求使用 Java 解决项目问题的朋友而言，本书也具有十分重要的参考价值。

结构安排

在本书中，你将经常看到如下几个标题："准备工作""操作步骤""工作原理""更多内容""另见"。

对于本书的每一节内容，我们会使用如下几个小标题来组织相关内容。

准备工作

本部分指出学习本节内容需要做的准备，也包含安装一些必需软件或做一些预先设置的内容。

操作步骤

本部分包含跟学中要做的具体步骤。

工作原理

本部分通常讲解与前一部分内容相关的更多细节。

更多内容

本部分讲解与前面内容相关的更多知识，通过阅读本部分内容，让读者掌握更多相关知识。

另见

本部分提供了一些有用的页面链接，从中读者可以获取更多与当前主题相关的有用内容。

本书使用说明

在本书中,在不同类型的信息之间,你将看到大量不同的文本类型。下面给出了这些类型的一些示例,并对它们所代表的含义进行了说明。

正文中出现的代码用语、数据表名、文件夹名、文件名、文件扩展名、路径名、虚拟URL、用户输入、推特标签显示如下:"在它们之间,你会发现一个名为 `lib` 的文件夹,它就是感兴趣的文件夹。"

代码块设置如下:

```
classVals = new ArrayList<String>();
  for (int i = 0; i < 5; i++){
     classVals.add("class" + (i + 1));
  }
```

命令行输入或输出写成如下形式:

```
@relation MyRelation

@attribute age numeric
@attribute name string
@attribute dob date yyyy-MM-dd
@attribute class {class1,class2,class3,class4,class5}

@data
35,'John Doe',1981-01-20,class3
30,'Harry Potter',1986-07-05,class1
```

正文中的新术语与关键词以粗体形式标识出来。你在屏幕截图中看到的词,比如在菜单或对话框中,出现在正文中的形式如下:"从 **Administration** 面板选择 **System info**"。

 警告或重要注释出现在这里。

 提示与技巧出现在这里。

资源与支持

本书由异步社区出品，社区（https://www.epubit.com/）为您提供相关资源和后续服务。

配套资源

本书提供如下资源：
- 本书源代码；
- 书中彩图文件。

要获得以上配套资源，请在异步社区本书页面中点击 配套资源 ，跳转到下载界面，按提示进行操作即可。注意：为保证购书读者的权益，该操作会给出相关提示，要求输入提取码进行验证。

如果您是教师，希望获得教学配套资源，请在社区本书页面中直接联系本书的责任编辑。

提交勘误

作者和编辑尽最大努力来确保书中内容的准确性，但难免会存在疏漏。欢迎您将发现的问题反馈给我们，帮助我们提升图书的质量。

当您发现错误时，请登录异步社区，按书名搜索，进入本书页面，点击"提交勘误"，输入勘误信息，单击"提交"按钮即可。本书的作者和编辑会对您提交的勘误进行审核，确认并接受后，您将获赠异步社区的 100 积分。积分可用于在异步社区兑换优惠券、样书或奖品。

扫码关注本书

扫描下方二维码,您将会在异步社区微信服务号中看到本书信息及相关的服务提示。

与我们联系

我们的联系邮箱是 contact@epubit.com.cn。

如果您对本书有任何疑问或建议,请您发邮件给我们,并请在邮件标题中注明本书书名,以便我们更高效地做出反馈。

如果您有兴趣出版图书、录制教学视频,或者参与图书翻译、技术审校等工作,可以发邮件给我们;有意出版图书的作者也可以到异步社区在线提交投稿(直接访问 www.epubit.com/selfpublish/submission 即可)。

如果您是学校、培训机构或企业,想批量购买本书或异步社区出版的其他图书,也可以发邮件给我们。

如果您在网上发现有针对异步社区出品图书的各种形式的盗版行为,包括对图书全部或部分内容的非授权传播,请您将怀疑有侵权行为的链接发邮件给我们。您的这一举动是对作者权益的保护,也是我们持续为您提供有价值的内容的动力之源。

关于异步社区和异步图书

"异步社区"是人民邮电出版社旗下 IT 专业图书社区,致力于出版精品 IT 技术图书和相关学习产品,为作译者提供优质出版服务。异步社区创办于 2015 年 8 月,提供大量精品 IT 技术图书和电子书,以及高品质技术文章和视频课程。更多详情请访问异步社区官网 https://www.epubit.com。

"异步图书"是由异步社区编辑团队策划出版的精品 IT 专业图书的品牌,依托于人民邮电出版社近 30 年的计算机图书出版积累和专业编辑团队,相关图书在封面上印有异步图书的 LOGO。异步图书的出版领域包括软件开发、大数据、AI、测试、前端、网络技术等。

异步社区

微信服务号

目录

第 1 章 获取数据与清洗数据 1
1.1 简介 2
1.2 使用 Java 从分层目录中提取所有文件名 3
 准备工作 3
 操作步骤 3
1.3 使用 Apache Commons IO 从多层目录中提取所有文件名 5
 准备工作 5
 操作步骤 5
1.4 使用 Java 8 从文本文件一次性读取所有内容 6
 操作步骤 7
1.5 使用 Apache Commons IO 从文本文件一次性读取所有内容 7
 准备工作 7
 操作方法 8
1.6 使用 Apache Tika 提取 PDF 文本 8
 准备知识 9
 操作步骤 9
1.7 使用正则表达式清洗 ASCII 文本文件 11
 操作步骤 11
1.8 使用 Univocity 解析 CSV 文件 12
 准备工作 13
 操作步骤 13
1.9 使用 Univocity 解析 TSV 文件 15
 准备工作 15
 操作步骤 16
1.10 使用 JDOM 解析 XML 文件 17
 准备工作 17
 操作步骤 18
1.11 使用 JSON.simple 编写 JSON 文件 20
 准备工作 20
 操作步骤 21
1.12 使用 JSON.simple 读取 JSON 文件 23
 准备工作 24

操作步骤 ························· 24
　1.13　使用 JSoup 从一个 URL
　　　　提取 Web 数据 ················· 26
　　　准备工作 ························· 26
　　　操作步骤 ························· 26
　1.14　使用 Selenium Webdriver
　　　　从网站提取 Web 数据 ·········· 29
　　　准备工作 ························· 29
　　　操作步骤 ························· 29
　1.15　从 MySQL 数据库读取表格
　　　　数据 ···························· 32
　　　准备工作 ························· 32
　　　操作步骤 ························· 32
第 2 章　为数据建立索引与搜索数据 ······ 35
　2.1　简介 ····························· 35
　2.2　使用 Apache Lucene 为数据
　　　　建立索引 ························· 35
　　　准备工作 ························· 36
　　　操作步骤 ························· 40
　　　工作原理 ························· 47
　2.3　使用 Apache Lucene 搜索带
　　　　索引的数据 ······················ 50
　　　准备工作 ························· 50
　　　操作步骤 ························· 51
第 3 章　数据统计分析 ······················· 56
　3.1　简介 ····························· 57
　3.2　生成描述性统计 ··············· 59
　　　操作步骤 ························· 59
　3.3　生成概要统计 ·················· 60
　　　操作步骤 ························· 60

　3.4　从多种分布生成概要统计 ······ 61
　　　操作步骤 ························· 62
　　　更多内容 ························· 63
　3.5　计算频率分布 ·················· 64
　　　操作步骤 ························· 64
　3.6　计算字符串中的词频 ········· 65
　　　操作步骤 ························· 65
　　　工作原理 ························· 67
　3.7　使用 Java 8 计算字符串中的
　　　　词频 ···························· 67
　　　操作步骤 ························· 67
　3.8　计算简单回归 ·················· 68
　　　操作步骤 ························· 69
　3.9　计算普通最小二乘回归 ······ 70
　　　操作步骤 ························· 70
　3.10　计算广义最小二乘回归 ····· 72
　　　操作步骤 ························· 72
　3.11　计算两组数据点的协方差 ···· 74
　　　操作步骤 ························· 74
　3.12　为两组数据点计算皮尔逊
　　　　相关系数 ······················ 75
　　　操作步骤 ························· 75
　3.13　执行配对 t 检验 ················ 76
　　　操作步骤 ························· 76
　3.14　执行卡方检验 ·················· 77
　　　操作步骤 ························· 78
　3.15　执行单因素方差分析
　　　　（one-way ANOVA test）········· 79
　　　操作步骤 ························· 79
　3.16　执行 K-S 检验 ················· 81
　　　操作步骤 ························· 81

第 4 章 数据学习 I ……………… 83

- 4.1 简介 ……………………………… 83
- 4.2 创建与保存 ARFF 文件 ……… 84
  - 操作步骤 ……………………… 87
- 4.3 对机器学习模型进行交叉验证 ……………………… 91
  - 操作步骤 ……………………… 91
- 4.4 对新的测试数据进行分类 …… 95
  - 准备工作 ……………………… 95
  - 操作步骤 ……………………… 96
- 4.5 使用过滤分类器对新测试数据分类 ………………… 102
  - 操作步骤 ……………………… 102
- 4.6 创建线性回归模型 …………… 105
  - 操作步骤 ……………………… 106
- 4.7 创建逻辑回归模型 …………… 108
  - 操作步骤 ……………………… 108
- 4.8 使用 K 均值算法对数据点进行聚类 ………………… 110
  - 操作步骤 ……………………… 110
- 4.9 依据类别对数据进行聚类处理 ……………………… 113
  - 操作方法 ……………………… 113
- 4.10 学习数据间的关联规则 …… 116
  - 准备工作 ……………………… 116
  - 操作步骤 ……………………… 116
- 4.11 使用低层方法、过滤方法、元分类器方法选择特征/属性 …………………… 118
  - 准备工作 ……………………… 119
  - 操作步骤 ……………………… 119

第 5 章 数据学习 II ……………… 125

- 5.1 简介 ……………………………… 125
- 5.2 使用 Java 机器学习库（Java-ML）向数据应用机器学习 …… 126
  - 准备工作 ……………………… 126
  - 操作步骤 ……………………… 128
- 5.3 使用斯坦福分类器对数据点分类 ……………………… 137
  - 准备工作 ……………………… 137
  - 操作步骤 ……………………… 140
  - 工作原理 ……………………… 141
- 5.4 使用 MOA 对数据点分类 …… 142
  - 准备工作 ……………………… 142
  - 操作步骤 ……………………… 144
- 5.5 使用 Mulan 对多标签数据点进行分类 ……………………… 147
  - 准备工作 ……………………… 147
  - 操作步骤 ……………………… 150

第 6 章 从文本数据提取信息 …… 154

- 6.1 简介 ……………………………… 154
- 6.2 使用 Java 检测标记（单词）… 155
  - 准备工作 ……………………… 155
  - 操作步骤 ……………………… 155
- 6.3 使用 Java 检测句子 …………… 160
  - 准备工作 ……………………… 160
  - 操作步骤 ……………………… 160
- 6.4 使用 OpenNLP 检测标记（单词）与句子 …………… 161

准备工作 ·· 162
操作步骤 ·· 163
6.5 使用 Stanford CoreNLP 从标记中提取词根、词性，以及识别命名实体 ········· 167
准备工作 ·· 167
操作步骤 ·· 169
6.6 使用 Java 8 借助余弦相似性测度测量文本相似度 ········· 171
准备工作 ·· 172
操作步骤 ·· 172
6.7 使用 Mallet 从文本文档提取主题 ············· 176
准备工作 ·· 177
操作步骤 ·· 179
6.8 使用 Mallet 对文本文档进行分类 ············· 184
准备工作 ·· 184
操作步骤 ·· 185
6.9 使用 Weka 对文本文档进行分类 ············· 189
准备工作 ·· 190
操作步骤 ·· 191

第 7 章 处理大数据 ·············· 194

7.1 简介 ·· 194
7.2 使用 Apache Mahout 训练在线逻辑回归模型 ······· 195
准备工作 ·· 195
操作步骤 ·· 198
7.3 使用 Apache Mahout 应用在线逻辑回归模型 ········· 202

准备工作 ·· 202
操作步骤 ·· 203
7.4 使用 Apache Spark 解决简单的文本挖掘问题 ········· 207
准备工作 ·· 208
操作步骤 ·· 210
7.5 使用 MLib 的 K 均值算法做聚类 ················· 214
准备工作 ·· 214
操作步骤 ·· 214
7.6 使用 MLib 创建线性回归模型 ············· 217
准备工作 ·· 217
操作步骤 ·· 218
7.7 使用 MLib 的随机森林模型对数据点进行分类 ········· 222
准备工作 ·· 222
操作步骤 ·· 223

第 8 章 数据深度学习 ·············· 229

8.1 简介 ·· 229
8.2 使用 DL4j 创建 Word2vec 神经网络 ············· 241
操作方法 ·· 241
工作原理 ·· 243
更多内容 ·· 246
8.3 使用 DL4j 创建深度信念神经网络 ············· 246
操作步骤 ·· 246
工作原理 ·· 250
8.4 使用 DL4j 创建深度自动编码器 ············· 254

操作步骤 ················· 254
　　　工作原理 ················· 256

第 9 章　数据可视化 ············· 259

9.1　简介 ···················· 259
9.2　绘制 2D 正弦曲线 ········· 260
　　　准备工作 ················· 260
　　　操作步骤 ················· 262
9.3　绘制直方图 ·············· 266
　　　准备工作 ················· 266
　　　操作步骤 ················· 268
9.4　绘制条形图 ·············· 273
　　　准备工作 ················· 274
　　　操作步骤 ················· 275

9.5　绘制箱线图或箱须图 ······· 279
　　　准备工作 ················· 279
　　　操作步骤 ················· 281
9.6　绘制散点图 ·············· 285
　　　准备工作 ················· 285
　　　操作步骤 ················· 286
9.7　绘制甜圈图 ·············· 289
　　　准备工作 ················· 289
　　　操作步骤 ················· 290
9.8　绘制面积图 ·············· 294
　　　准备工作 ················· 294
　　　操作步骤 ················· 295

第 1 章
获取数据与清洗数据

本章涵盖如下内容：

- 使用 Java 从分层目录中提取所有文件名；
- 使用 Apache Commons IO 从分层目录中提取所有文件名；
- 使用 Java 8 同时从多个文本文件读取内容；
- 使用 Apache Commons IO 同时从多个文本文件读取内容；
- 使用 Apache Tika 提取 PDF 文本；
- 使用正则表达式清洗 ASCII 文本文件；
- 使用 Univocity 解析 CSV（逗号分隔）文件；
- 使用 Univocity 解析 TSV（制表符分隔）文件；
- 使用 JDOM 解析 XML 文件；
- 使用 JSON.simple 编写 JSON 文件；
- 使用 JSON.simple 读取 JSON 文件；
- 使用 JSoup 从一个 URL 地址提取 Web 数据；
- 使用 Selenium Webdriver 从一个网站提取 Web 数据；
- 从 MYSQL 数据库读取表格数据。

1.1 简介

每个数据科学家都需要处理存储在磁盘中的数据,这些数据涉及的格式有 ASCII 文本、PDF、XML、JSON 等。此外,数据还可以存储在数据库表格中。在对数据进行分析之前,数据科学家首先要做的是从这些数据源获取各种格式的数据,并对这些数据进行清洗,去除其中的噪声。本章我们将学习这些内容,即了解如何从不同数据源获取各种格式的数据。

在这一过程中,我们将用到外部 Java 库(Java 归档文件,简称 JAR 文件),这些库的使用不仅限于本章,还贯穿于整本书。这些库由不同开发者或组织开发,方便了大家的使用。编写代码时,我们会用到 Eclipse IDE 工具,它是 Windows 平台下最好的集成开发环境,全书都会使用它。接下来,我们将讲解如何导入任意一个外部 JAR 文件,以下各个部分将指导你把外部 JAR 文件导入到项目中,跟随步骤动手去做即可。

对于一个 Eclipse 项目,你可以采用如下方法添加 JAR 文件:首先依次单击 "**Project|Build Path|Configure Build Path**",在 **Libraries** 选项卡中,单击 "**Add External JARs**",选择你想添加到项目的外部 JAR 文件,如图 1-1 所示。

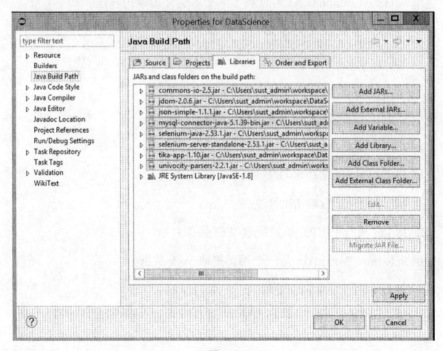

图 1-1

1.2 使用 Java 从分层目录中提取所有文件名

这部分内容(以及后面各部分内容)是为那些想从复杂目录结构中提取文件路径与名称的数据科学家准备的,以方便进一步进行后续分析。这里的复杂目录结构是指在一个根目录下包含大量目录与文件。

准备工作

开始之前,需要做如下准备工作。

1. 创建复杂的目录结构(目录层数你自己决定)。
2. 在其中一些目录中创建文本文件,而在另一些目录中留空。

操作步骤

1. 首先编写一个 static 方法,即 listFiles(File rootDir),它带有一个 File 类型的参数,该参数可以是根目录或起始目录。这个方法将返回一系列文件,这些文件存在于参数所指定的根目录(以及其他所有下级子目录)中。

   ```
   public static Set<File> listFiles(File rootDir) {
   ```

2. 然后,创建一个 HashSet 对象,用来包含文件信息。

   ```
   Set<File> fileSet = new HashSet<File>();
   ```

3. 在创建好 HashSet 对象之后,要检查参数指定的根目录及其子目录是否为 null。当为 null 时,直接把 HashSet 对象返回即可,不需要进行进一步处理。

   ```
   if (rootDir == null || rootDir.listFiles() == null){
               return fileSet;
       }
   ```

4. 接着,检查根目录中的每个目录(或文件),判断它是文件还是目录。如果是文件,就把它添加到 HashSet 中;如果是一个目录,就递归地调用本方法,并把当前目录路径与名称传递给它。

   ```
   for (File fileOrDir : rootDir.listFiles()) {
   ```

```
            if (fileOrDir.isFile()){
              fileSet.add(fileOrDir);
            }
            else{
              fileSet.addAll(listFiles(fileOrDir));
            }
        }
```

5. 最后,把 HashSet 返回给该方法的调用者。

```
        return fileSet;
      }
```

listFiles(File rootDir)方法的完整代码如下,包含执行该方法所需要的类与驱动方法。

```
import java.io.File;
import java.util.HashSet;
import java.util.Set;

public class TestRecursiveDirectoryTraversal {
    public static void main(String[] args){
        System.out.println(listFiles(new File("Path for root
            directory")).size());
    }
    public static Set<File> listFiles(File rootDir) {
        Set<File> fileSet = new HashSet<File>();
        if(rootDir == null || rootDir.listFiles()==null){
            return fileSet;
        }
        for (File fileOrDir : rootDir.listFiles()) {
              if (fileOrDir.isFile()){
                fileSet.add(fileOrDir);
              }
              else{
                fileSet.addAll(listFiles(fileOrDir));
              }
        }
        return fileSet;
    }
}
```

请注意，代码中的 HashSet 用来存储文件路径与名称。这意味着我们不会有任何重复项，这是因为 Java 中的 Set 这种数据结构不包含重复项。

1.3 使用 Apache Commons IO 从多层目录中提取所有文件名

你可以使用前面一部分演示的操作步骤，采用递归方法把多层目录中的文件名列出来。除此之外，我们还有另外一种更简单、更方便的方法来完成它，那就是使用 Apache Commons IO，并且只需编写少量代码即可。

准备工作

开始之前，需要做如下准备。

1. 本部分会用到一个名称为 Commons IO 的 Java 库，它来自于 Apache 基金会。全书中，我们会使用 Commons IO 2.5 版本，请从 Commons 官网下载 JAR 文件。
2. 在 Eclipse 中，把下载的 JAR 文件包含到你的项目中（作为外部 JAR 文件）。

操作步骤

1. 创建 listFiles 方法，它带有一个参数，用来指定层级目录的根目录。

    ```
    public void listFiles(String rootDir){
    ```

2. 创建一个文件对象，并把根目录名传递给它。

    ```
    File dir = new File(rootDir);
    ```

3. Apache Commons 库的 FileUtils 类中包含一个名称为 listFiles() 方法。使用这个方法提取所有文件名，并且把它们放入一个带有 <File> 泛型的列表变量中。使用 TrueFileFilter.INSTANCE 来匹配所有目录。

    ```
    List<File> files = (List<File>) FileUtils.listFiles(dir,
      TrueFileFilter.INSTANCE, TrueFileFilter.INSTANCE);
    ```

4. 我们可以像下面这样把文件名显示在标准输出中。由于我们把文件名放入了一个列表之中，所以我们可以通过某种方法对这些文件中的数据进行进一步处理。

```
    for (File file : files) {
        System.out.println("file: " + file.getAbsolutePath());
    }
```

5. 关闭方法。

```
}
```

完整代码包括方法代码、类代码，以及驱动方法，如下所示：

```
import java.io.File;
import java.util.List;
import org.apache.commons.io.FileUtils;
import org.apache.commons.io.filefilter.TrueFileFilter;

public class FileListing{
    public static void main (String[] args){
        FileListing fileListing = new FileListing();
        fileListing.listFiles("Path for the root directory here");
    }
    public void listFiles(String rootDir){
        File dir = new File(rootDir);

        List<File> files = (List<File>) FileUtils.listFiles(dir,
          TrueFileFilter.INSTANCE, TrueFileFilter.INSTANCE);
        for (File file : files) {
            System.out.println("file: " + file.getAbsolutePath());
        }
    }
}
```

如果你想把带有一些特定扩展名的文件列出来，还可以使用 Apache Commons 库中的 `listFiles` 方法。但是这个方法的参数有些不同，它拥有 3 个参数，分别为文件目录、扩展名（`String[]`）、递归与否。在这个库中还有一个有趣的方法，即 listFilesAndDirs（File directory, IOFileFilter fileFilter, IOFileFilter dirFilter），如果你想把文件与目录全部列出来，可以使用它。

1.4 使用 Java 8 从文本文件一次性读取所有内容

在许多场合下，数据科学家所拥有的数据是文本格式的。我们有很多方法可以用来读取文本文件的内容，这些方法各具优缺点：一些方法执行起来耗时、耗内存，而另一些方法执行速度很快，也不需要消耗太多计算机内存；一些方法可以把全部文本内容一次性读出，而另一些方法则只能一行行地读取文本文件。至于到底要选择哪种方法，则取决于你所面对的任务，以及你决定采用何种方法来处理这个任务。

在这部分中，我们将演示使用 Java 8 把文本文件的全部内容一次性读出来的方法。

操作步骤

1. 首先，创建一个 String 对象，用来保存待读取的文本文件的目录与名称。

   ```
   String file = "C:/dummy.txt";
   ```

2. 使用 Paths 类的 get()方法，可以得到待读文件的路径。get()方法的参数是 String 对象，用来指定文件名，它的输出作为 lines()方法的输入。lines() 方法包含于 Files 类之中，用来读取一个文件的所有行，并且返回 Stream，也就是说，这个方法的输出定向到一个 Stream 变量。因为我们的 dummy.txt 文件中包含字符串数据，所以把 Stream 变量的泛型设置为 String。

整个读取过程需要放入一个 try...catch 块中，用来应对读取过程中可能发生的异常，比如当试图读取的文件不存在或已损坏时，就会抛出异常。

下面代码用来把 dummy.txt 文件中的内容全部显示出来。在 stream 变量中包含着文本文件的所有行，所以需要使用它的 forEach()方法显示出每行内容。

```
try (Stream<String> stream = Files.lines(Paths.get(file))) {
stream.forEach(System.out::println); } catch (IOException e) {
System.out.println("Error reading " + file.getAbsolutePath());
}
```

1.5 使用 Apache Commons IO 从文本文件一次性读取所有内容

在上一节中我们学习了使用 Java8 从文本文件中一次性读取所有内容，其实我们也可以使用 Apache Commons IO API 一次性读取文本文件的所有内容。

准备工作

开始之前，需要做如下准备。

1. 本部分，我们会用到一个名为 Apache Commons IO 的 Java 库。
2. 在 Eclipse 中，把下载好的 JAR 文件包含到你的项目中。

操作方法

1. 假设你要读取的文件为 dummy.txt，它位于 C:/ 目录之下。首先，需要创建一个文件对象，用来访问这个文件，如下所示：

   ```
   File file = new File("C:/dummy.txt");
   ```

2. 接着，创建一个字符串对象，用来保存文件中的文本内容。这里我们要使用 `readFileToString()` 方法，它来自于 Apache Commons IO 库，是 `FileUtils` 类的一个成员方法。调用这个方法的方式有很多，但是现在，你只需知道我们要传递两个参数给它，第一个参数是 `file` 对象，用来指定要读取的文件，第二个参数是文件的编码，在示例中，我们将其设置为 UTF-8。

   ```
   String text = FileUtils.readFileToString(file, "UTF-8");
   ```

3. 只要使用上面两行代码，我们就可以读取文本文件内容，并将它们存入一个 String 变量中。但是，你可不是一个普通的数据科学家，你比其他人要聪明得多。所以，你在上面两行代码的前后又添加了几行代码，用来处理 Java 方法抛出的异常，比如你试图读取的文件不存在或者已经损坏，就会触发异常。为此，我们需要把上面两行代码放入到一个 `try...catch` 块之中，如下所示：

   ```
   File file = new File("C:/dummy.txt");
   try {
   String text = FileUtils.readFileToString(file, "UTF-8");
   } catch (IOException e) {
   System.out.println("Error reading " + file.getAbsolutePath());
   }
   ```

1.6 使用 Apache Tika 提取 PDF 文本

在解析与提取数据时，最难搞的文件类型之一是 PDF 文件。有些 PDF 文件甚至无法解析，因为它们有密码保护，而其他一些则包含着扫描的文本与图像。所以，这种动态文件类型有时会成为数据科学家的梦魇。本部分演示如何使用 Apache Tika 从 PDF 文件提取文本，当然前提是 PDF 文件没有被加密，也没有密码保护，而只包含非扫描的文本。

准备知识

开始之前，需要先做如下准备。

1. 下载 Apache Tika 1.10 JAR 文件，并且将其作为外部 Java 库包含到你的 Eclipse 项目中。
2. 把任意一个未锁定的 PDF 文件保存到 C:/目录之下，并且命名为 `testPDF.pdf`。

操作步骤

1. 创建一个名称为 `convertPdf(String)` 的方法，它带有一个字符串参数，用来指定 PDF 文件名称。

   ```
   public void convertPDF(String fileName){
   ```

2. 创建一个输入流，用来以字节流的形式包含 PDF 数据。

   ```
   InputStream stream = null;
   ```

3. 创建一个 `try` 块，如下所示：

   ```
   try{
   ```

4. 把文件指派给刚刚创建好的 `stream`。

   ```
   stream = new FileInputStream(fileName);
   ```

5. 在 Apache Tika 包中包含着许多不同的解析器。如果你不知道该选用哪一个，或者说你还有其他类型的文档需要转换，那么你应该使用 `AutoDetectParser` 解析器，如下所示：

   ```
   AutoDetectParser parser = new AutoDetectParser();
   ```

6. 创建一个 handler，用来处理文件的正文内容。请注意，创建时需要把构造函数的参数设为 `-1`。通常，Apache Tika 会对处理的文件进行限制，要求它至多包含 100 000

个字符。使用-1 让这个 handler 忽略这个限制。

```
BodyContentHandler handler = new BodyContentHandler(-1);
```

7. 创建一个 metadata 对象。

```
Metadata metadata = new Metadata();
```

8. 调用解析器对象的 parser() 方法,并把上面创建的这些对象传递给它。

```
parser.parse(stream, handler, metadata, new ParseContext());
```

9. 使用 handler 对象的 tostring() 方法,获取从文件中提取的正文文本。

```
System.out.println(handler.toString());
```

10. 关闭 try 块,并且添加 catch 与 finally 块。最后,关闭整个方法,如下所示:

```
}catch (Exception e) {
        e.printStackTrace();
    }finally {
        if (stream != null)
            try {
                stream.close();
            } catch (IOException e) {
                System.out.println("Error closing stream");
            }
    }
}
```

下面代码包含 convertPdf(String)方法的完整代码,以及相应的类与驱动方法。在调用 convertPdf(String)方法时,你需要提供待转换的 PDF 文件的路径与名称,即把该方法的参数指定为 C:/testPDF.pdf。

```
import java.io.FileInputStream;
import java.io.IOException;
import java.io.InputStream;
import org.apache.tika.metadata.Metadata;
import org.apache.tika.parser.AutoDetectParser;
import org.apache.tika.parser.ParseContext;
import org.apache.tika.sax.BodyContentHandler;
```

```
public class TestTika {
    public static void main(String args[]) throws Exception {
        TestTika tika = new TestTika();
        tika.convertPdf("C:/testPDF.pdf");
    }
    public void convertPdf(String fileName){
        InputStream stream = null;
        try {
            stream = new FileInputStream(fileName);
            AutoDetectParser parser = new AutoDetectParser();
            BodyContentHandler handler = new BodyContentHandler(-1);
            Metadata metadata = new Metadata();
            parser.parse(stream, handler, metadata, new
                ParseContext());
            System.out.println(handler.toString());
        }catch (Exception e) {
            e.printStackTrace();
        }finally {
            if (stream != null)
                try {
                    stream.close();
                } catch (IOException e) {
                    System.out.println("Error closing stream");
                }
        }
    }
}
```

1.7 使用正则表达式清洗 ASCII 文本文件

ASCII 文本文件中通常会包含一些非必要的字符，这些字符通常产生于转换过程中，比如把 PDF 转换为文本或把 HTML 转换为文本的过程中。并且，这些字符常常被看作噪声，它们是数据处理的主要障碍之一。本部分，我们学习使用正则表达式为 ASCII 文本数据清洗一些噪声的方法。

操作步骤

1. 创建一个名为 cleanText(String) 的方法，它带有一个 String 类型的参数，用来指定要清洗的文本。

```
        public String cleanText(String text){
```

2. 在你的方法中，添加如下几行代码，而后把清洗后的文本返回，并关闭方法。在如下代码中，第一行代码用来去掉非 ASCII 字符，紧接的一行用来把连续的空格字符替换为单个空格字符。第三行用来清除所有 ASCII 控制字符。第四行用来去除 ASCII 非打印字符。最后一行用来从 Unicode 移除非打印字符。

```
        text = text.replaceAll("[^p{ASCII}]","");
        text = text.replaceAll("s+", " ");
        text = text.replaceAll("p{Cntrl}", "");
        text = text.replaceAll("[^p{Print}]", "");
        text = text.replaceAll("p{C}", "");

        return text;
        }
```

以下代码是方法的完整代码，包含相应类与驱动方法。

```
public class CleaningData {
   public static void main(String[] args) throws Exception {
      CleaningData clean = new CleaningData();
      String text = "Your text here you have got from some file";
      String cleanedText = clean.cleanText(text);
      //清洗文本处理
   }
   public String cleanText(String text){
      text = text.replaceAll("[^p{ASCII}]","");
        text = text.replaceAll("s+", " ");
        text = text.replaceAll("p{Cntrl}", "");
        text = text.replaceAll("[^p{Print}]", "");
        text = text.replaceAll("p{C}", "");
        return text;
   }
}
```

1.8 使用 Univocity 解析 CSV 文件

对数据科学家来说，另一种经常处理的文件格式是 CSV（逗号分隔）文件，在这种文件中数据之间通过逗号进行分隔。CSV 文件非常流行，因为大部分电子表格应用程序都可以读取它，比如 MS Excel。

本部分，我们将学习解析 CSV 文件，以及处理所提取的数据点的方法。

准备工作

开始之前，需要先做如下准备。

1. 下载 Univocity JAR 文件，并将其作为外部库添加到你的 Eclipse 项目中。
2. 使用 Notepad 创建一个 CSV 文件，它包含如下数据。创建好之后，把文件的扩展名修改为 .csv，并把它保存到 C 盘之下，即 C:/testCSV.csv。

   ```
   Year,Make,Model,Description,Price
   1997,Ford,E350,"ac, abs, moon",3000.00
   1999,Chevy,"Venture ""Extended Edition""","",4900.00
   1996,Jeep,Grand Cherokee,"MUST SELL!
   air, moon roof, loaded",4799.00
   1999,Chevy,"Venture ""Extended Edition, Very Large""",,5000.00
   ,,"Venture ""Extended Edition""","",4900.00
   ```

操作步骤

1. 创建一个名为 `parseCsv(String)` 的方法，它带有一个 String 类型的参数，用来指定待解析的文件名。

   ```
   public void parseCsv(String fileName){
   ```

2. 而后创建一个配置对象，该对象用来提供多种配置选项。

   ```
   CsvParserSettings parserSettings = new CsvParserSettings();
   ```

3. 借助于配置对象，你可以打开解析器的自动检测功能，让它自动侦测输入中包含何种行分隔符序列。

   ```
   parserSettings.setLineSeparatorDetectionEnabled(true);
   ```

4. 创建一个 RowListProcessor 对象，用来把每个解析的行存储在列表中。

   ```
   RowListProcessor rowProcessor = new RowListProcessor();
   ```

5. 你可以使用 RowProcessor 来配置解析器，以对每个解析行的值进行处理。你可以在 `com.univocity.parsers.common.processor` 包中找到更多 RowProcessors，

但是你也可以自己创建。

```
parserSettings.setRowProcessor(rowProcessor);
```

6. 如果待解析的 CSV 文件包含标题头，你可以把第一个解析行看作文件中每个列的标题。

```
parserSettings.setHeaderExtractionEnabled(true);
```

7. 接下来，使用给定的配置创建一个 parser 实例。

```
CsvParser parser = new CsvParser(parserSettings);
```

8. parser 实例的 parse() 方法用来解析文件，并把每个经过解析的行指定给前面定义的 RowProcessor。

```
parser.parse(new File(fileName));
```

9. 如果解析中包含标题，则可使用如下代码获取这些标题。

```
String[] headers = rowProcessor.getHeaders();
```

10. 随后，你可以很容易地处理这个字符串数组，以获取这些标题值。

11. 另一方面，我们在列表中可以找到行值。只要使用一个 for 循环即可把列表打印出来，如下所示。

```
List<String[]> rows = rowProcessor.getRows();
for (int i = 0; i < rows.size(); i++){
    System.out.println(Arrays.asList(rows.get(i)));
}
```

12. 最后，关闭方法。

```
}
```

整个方法的完整代码如下所示：

```
import java.io.File;
import java.util.Arrays;
import java.util.List;
```

```java
import com.univocity.parsers.common.processor.RowListProcessor;
import com.univocity.parsers.csv.CsvParser;
import com.univocity.parsers.csv.CsvParserSettings;

public class TestUnivocity {
    public void parseCSV(String fileName){
        CsvParserSettings parserSettings = new CsvParserSettings();
        parserSettings.setLineSeparatorDetectionEnabled(true);
        RowListProcessor rowProcessor = new RowListProcessor();
        parserSettings.setRowProcessor(rowProcessor);
        parserSettings.setHeaderExtractionEnabled(true);
        CsvParser parser = new CsvParser(parserSettings);
        parser.parse(new File(fileName));

        String[] headers = rowProcessor.getHeaders();
        List<String[]> rows = rowProcessor.getRows();
        for (int i = 0; i < rows.size(); i++){
           System.out.println(Arrays.asList(rows.get(i)));
        }
    }
    public static void main(String[] args){
       TestUnivocity test = new TestUnivocity();
       test.parseCSV("C:/testCSV.csv");
    }
}
```

> 有很多采用 Java 编写的 CSV 解析器。但是，相比较而言，Univocity 是执行速度最快的一个。

1.9 使用 Univocity 解析 TSV 文件

不同于 CSV 文件，TSV（制表符分隔）文件中所包含的数据通过 TAB 制表符进行分隔。本部分，我们将学习使用 Univocity 从 TSV 文件提取数据点的方法。

准备工作

开始之前，先做如下准备工作。

1. 下载 Univocity JAR 文件，并将其作为外部库包含到你的 Eclipse 项目中。

2. 使用 Notepad 创建一个 TSV 文件，它包含如下数据。创建好之后，把文件的扩展名修改为 .tsv，并把它保存到 C 盘之下，即 C:/testTSV.tsv。

```
Year     Make      Model     Description     Price
1997     Ford      E350      ac, abs, moon   3000.00
1999     Chevy     Venture   "Extended Edition"     4900.00
1996     Jeep      Grand Cherokee   MUST SELL!nair, moon roof, loaded 4799.00
1999     Chevy     Venture   "Extended Edition, Very Large"     5000.00
                   Venture   "Extended Edition"     4900.00
```

操作步骤

1. 创建一个名称为 parseTsv(String) 的方法，它带有一个 String 类型的参数，用来指定待解析的文件名。

   ```
   public void parseTsv(String fileName){
   ```

2. 本部分中 TSV 文件的行分隔符为换行符或 n。为了把字符 n 设置为行分隔符，修改设置如下所示。

   ```
   settings.getFormat().setLineSeparator("n");
   ```

3. 使用这些设置，创建一个 TSV 解析器。

   ```
   TsvParser parser = new TsvParser(settings);
   ```

4. 使用如下代码，把 TSV 文件中的所有行一次性解析出来。

   ```
   List<String[]> allRows = parser.parseAll(new File(fileName));
   ```

5. 遍历列表对象，打印或处理数据行，代码如下所示。

   ```
   for (int i = 0; i < allRows.size(); i++){
        System.out.println(Arrays.asList(allRows.get(i)));
   {
   ```

6. 最后，关闭方法。

   ```
   }
   ```

下面代码中包含整个方法的完整代码，以及相应类与驱动方法。

```
import java.io.File;
import java.util.Arrays;
import java.util.List;

import com.univocity.parsers.tsv.TsvParser;
import com.univocity.parsers.tsv.TsvParserSettings;

public class TestTsv {
   public void parseTsv(String fileName){
      TsvParserSettings settings = new TsvParserSettings();
      settings.getFormat().setLineSeparator("n");
      TsvParser parser = new TsvParser(settings);
      List<String[]> allRows = parser.parseAll(new File(fileName));
      for (int i = 0; i < allRows.size(); i++){
         System.out.println(Arrays.asList(allRows.get(i)));
      }
   }
}
```

1.10 使用 JDOM 解析 XML 文件

通常，文本数据是没有结构的，不同于文本数据，在 XML 文件中的数据是具有结构的。XML 是组织数据的一种流行方式，借助它，我们可以非常方便地准备、传递以及利用数据。有很多方法可以用来解析 XML 文件的内容。本书中，我们将学习使用一个名为 JDOM 的 Java 库来解析 XML 文件。

准备工作

开始之前，先做如下准备工作。

1. 下载 JDOM 2.06 版本（JAR 文件）。

2. 在 Eclipse 中，创建一个项目，并把上面下载的 JAR 文件作为外部 JAR 文件包含进去。

3. 打开 notepad，新建一个名称为 `dummyxml` 的文件，文件扩展名为 `.xml`，文件包含的内容简单如下：

   ```
   <?xml version="1.0"?>
   ```

```
<book>
    <author>
        <firstname>Alice</firstname>
        <lastname>Peterson</lastname>
    </author>
    <author>
        <firstname>John</firstname>
        <lastname>Doe</lastname>
    </author>
</book>
```

操作步骤

1. 创建一个名称为 `builder` 的 `SAXBuilder` 对象。

   ```
   SAXBuilder builder = new SAXBuilder();
   ```

2. 接下来，你需要创建一个 `File` 对象，用来指向待解析的 XML 文件。如果你已经把 XML 文件保存到 C 盘之下，则将其放入如下代码片段中。

   ```
   File file = new File("c:/dummyxml.xml");
   ```

3. 在 `try` 语句块中，需要创建一个 `Document` 对象，它表示你的 XML 文件。

   ```
   try {
       Document document = (Document) builder.build(file);
   ```

4. 在解析呈现树状结构的 XML 文件时，需要知道文件的根元素，以便开始遍历整个树（换言之，开始进行系统的解析）。因此，需要创建一个 `Element` 类型的 `rootNode` 对象，用来保存根元素，在我们的示例中，它对应于 `<book>` 节点。

   ```
   Element rootNode = document.getRootElement();
   ```

5. 接着，获取根节点下所有名称为 `author` 的子节点。由于调用 `getChildren()` 方法所得到的是子节点列表，所以还需要有一个列表变量来存储它们。

   ```
   List list = rootNode.getChildren("author");
   ```

6. 然后，使用 `for` 循环遍历整个子节点列表，以获取列表中的项目元素。每个元素都存储在 `Element` 类型的 `node` 变量中。这个变量有一个名称为 `getChildText()` 的方法，其参数为子元素名称，返回子元素的文本内容，如果指定的子元素不存在，就返回 null。这个方法用起来非常方便，因为调用 `getChild().getText()` 可能会抛出 NullPointerException 异常。

```
for (int i = 0; i < list.size(); i++) {
    Element node = (Element) list.get(i);
System.out.println("First Name : " +
  node.getChildText("firstname"));
System.out.println("Last Name : " +
  node.getChildText("lastname"));
}
```

7. 最后，关闭 try 语句块，并添加如下 catch 语句块处理可能遇到的异常。

```
} catch (IOException io) {
    System.out.println(io.getMessage());
} catch (JDOMException jdomex) {
    System.out.println(jdomex.getMessage());
}
```

完整代码如下：

```
import java.io.File;
import java.io.IOException;
import java.util.List;

import org.jdom2.Document;
import org.jdom2.Element;
import org.jdom2.JDOMException;
import org.jdom2.input.SAXBuilder;

public class TestJdom {

  public static void main(String[] args){
     TestJdom test = new TestJdom();
     test.parseXml("C:/dummyxml.com");
  }
  public void parseXml(String fileName){
     SAXBuilder builder = new SAXBuilder();
     File file = new File(fileName);
     try {
        Document document = (Document) builder.build(file);
```

```
            Element rootNode = document.getRootElement();
            List list = rootNode.getChildren("author");
            for (int i = 0; i < list.size(); i++) {
               Element node = (Element) list.get(i);
               System.out.println("First Name : " +
                   node.getChildText("firstname"));
               System.out.println("Last Name : " +
                   node.getChildText("lastname"));
            }
        } catch (IOException io) {
          System.out.println(io.getMessage());
        } catch (JDOMException jdomex) {
          System.out.println(jdomex.getMessage());
        }
     }
}
```

> XML 解析器类型多样，每种解析器都各有优点。**Dom Parser**。这种解析器会把文档的完整内容加载到内存中，并在内存中创建自己的层次树。**SAX Parser**：这种解析器不会把整个文档全部加载到内存中，文档的解析基于事件触发。**JDOM Parser**：JDOM 解析器采用类似 DOM 解析器的方式解析文档，但是更加便捷。**StAX Parser**：这种解析器采用类似于 SAX 解析器的方式处理文档，但是效率更高。**XPath Parser**：这类解析器基于路径表达式来解析文档，经常与 XSLT 一起使用。**DOM4J Parser**：这是一个使用 Java 集合框架（该框架提供了对 DOM、SAX、JAXP 的支持）来解析 XML、XPath、XSLT 的 Java 库。

1.11　使用 JSON.simple 编写 JSON 文件

类似于 XML 文件，JSON 也是一种人类可读的轻量级数据交换格式。JSON 是 JavaScript Object Notation（JavaScript 对象表示法）的首字母缩写。JSON 正成为一种由现代 Web 应用程序所生成与解析的流行格式。本部分中，我们将学习如何编写 JSON 文件。

准备工作

开始之前，先做如下准备工作。

下载 json-simple-1.1.1.jar，并将其作为外部库添加到你的 Eclipse 项目中。

操作步骤

1. 创建一个名称为 `writeJson(String outFileName)` 的方法，它带有一个 String 类型参数，用来指定要生成的 JSON 文件名。本节我们将创建它，并与 JSON 信息一起输出。

2. 创建一个 JSON 对象，使用它的 `put()` 方法，添加几个字段，比如图书与作者。下面代码用来创建一个 JSON 对象，然后添加书名及其作者名字。

   ```
   JSONObject obj = new JSONObject();
       obj.put("book", "Harry Potter and the Philosopher's Stone");
       obj.put("author", "J. K. Rowling");
   ```

3. 接着，假设针对这本书有 3 个书评。我们可以把它们一起放入一个 JSON 数组中，具体操作如下：首先，我们使用数组对象的 `add()` 方法添加评论，当把所有评论添加到数组之后，我们再把数组添加到上一步所创建的 JSON 对象。

```
JSONArray list = new JSONArray();

list.add("There are characters in this book that will remind us of all the
people we have met. Everybody knows or knew a spoilt, overweight boy like
Dudley or a bossy and interfering (yet kind-hearted) girl like Hermione");
list.add("Hogwarts is a truly magical place, not only in the most obvious
way but also in all the detail that the author has gone to describe it so
vibrantly.");
list.add("Parents need to know that this thrill-a-minute story, the first
in the Harry Potter series, respects kids' intelligence and motivates them
to tackle its greater length and complexity, play imaginative games, and
try to solve its logic puzzles. ");

obj.put("messages", list);
```

4. 接下来，我们把 JSON 对象中的信息写入到一个输出文件中，后面我们将使用这个文件演示如何读取与解析 JSON 文件。下面的 `try...catch` 代码块用来把信息写到 JSON 文件。

   ```
   try {
           FileWriter file = new FileWriter("c:test.json");
           file.write(obj.toJSONString());
           file.flush();
   ```

```
            file.close();

    } catch (IOException e) {
            //your message for exception goes here.
    }
```

5. 我们也可以把 JSON 对象的内容显示在标准的输出中,如下所示。

   ```
   System.out.print(obj);
   ```

6. 最后,关闭方法。

   ```
   }
   ```

完整代码如下,包括 writeJson()方法的全部代码,以及相应类与用于调用 writeJson() 方法的驱动方法(main 方法),代码执行后会生成一个 JSON 文件。

```
import java.io.FileWriter;
import java.io.IOException;
import org.json.simple.JSONArray;
import org.json.simple.JSONObject;

public class JsonWriting {

    public static void main(String[] args) {
        JsonWriting jsonWriting = new JsonWriting();
        jsonWriting.writeJson("C:/testJSON.json");
    }

    public void writeJson(String outFileName){
        JSONObject obj = new JSONObject();
        obj.put("book", "Harry Potter and the Philosopher's Stone");
        obj.put("author", "J. K. Rowling");

        JSONArray list = new JSONArray();
        list.add("There are characters in this book that will remind us
            of all the people we have met. Everybody knows or knew a
            spoilt, overweight boy like Dudley or a bossy and interfering
                (yet kind-hearted) girl like Hermione");
        list.add("Hogwarts is a truly magical place, not only in the most
            obvious way but also in all the detail that the author has gone
            to describe it so vibrantly.");
        list.add("Parents need to know that this thrill-a-minute story,
            the first in the Harry Potter series, respects kids'
```

```java
            intelligence and motivates them to tackle its greater length
            and complexity, play imaginative games, and try to solve
            its logic puzzles. ");

    obj.put("messages", list);

    try {

        FileWriter file = new FileWriter(outFileName);
        file.write(obj.toJSONString());
        file.flush();
        file.close();

    } catch (IOException e) {
        e.printStackTrace();
    }

    System.out.print(obj);
    }
}
```

在输出文件中将包含如下数据。请注意，这里为了增加可读性，我们对输出的结果进行了一些调整，实际输出的是一大段纯文本。

```
{
"author":"J. K. Rowling",
"book":"Harry Potter and the Philosopher's Stone",
"messages":[
        "There are characters in this book that will remind us of all the
people we have met. Everybody knows or knew a spoilt, overweight boy like
Dudley or a bossy and interfering (yet kind-hearted) girl like Hermione",
        "Hogwarts is a truly magical place, not only in the most obvious
way but also in all the detail that the author has gone to describe it so
vibrantly.",
        "Parents need to know that this thrill-a-minute story, the first
in the Harry Potter series, respects kids' intelligence and motivates them
to tackle its greater length and complexity, play imaginative games, and
try to solve its logic puzzles."
        ]
}
```

1.12　使用 JSON.simple 读取 JSON 文件

在本部分，我们将学习读取或解析 JSON 文件的方法。在这里，我们将使用上一小节创建的 JSON 文件作为待读取或解析的文件。

准备工作

开始之前,先做如下准备。

根据上一小节的操作步骤,创建一个 JSON 文件,里面包含图书名称、作者与书评信息。我们将使用这个文件来学习解析与读取 JSON 文件的方法。

操作步骤

1. 在读取或解析 JSON 文件之前,我们需要先创建一个 JSON 解析器。

   ```
   JSONParser parser = new JSONParser();
   ```

2. 接着,在 try 语句块中,把书名与作者字段的值提取出来。在此之前,需要先使用解析器的 parse() 方法读取 JSON 文件。parse() 方法会把文件内容作为一个 Object 返回。因此,我们需要一个 Object 类型的变量来保存文件内容。然后,把 object 指派给一个 JSON 对象,以便进行进一步处理。请注意,在把 object 指派给 JSON 对象时,需要进行类型转换处理。

   ```
   try {
     Object obj = parser.parse(new FileReader("c:test.json"));
     JSONObject jsonObject = (JSONObject) obj;

     String name = (String) jsonObject.get("book");
     System.out.println(name);

     String author = (String) jsonObject.get("author");
     System.out.println(author);
   }
   ```

3. 接下来需要从 JSON 对象提取的字段是评论字段,它是一个数组。我们需要使用如下代码对数组进行遍历。

   ```
   JSONArray reviews = (JSONArray) jsonObject.get("messages");
   Iterator<String> iterator = reviews.iterator();
   while (iterator.hasNext()) {
      System.out.println(iterator.next());
   }
   ```

4. 最后,添加 catch 语句块,处理解析期间可能遇到的 3 种异常,然后关闭方法。

```
        } catch (FileNotFoundException e) {
            //你的异常处理代码
        } catch (IOException e) {
            //你的异常处理代码
        } catch (ParseException e) {
            //你的异常处理代码
        }
    }
```

完整代码整理如下，包括完整的方法代码、相应类，以及驱动方法。

```
import java.io.FileNotFoundException;
import java.io.FileReader;
import java.io.IOException;
import java.util.Iterator;
import org.json.simple.JSONArray;
import org.json.simple.JSONObject;
import org.json.simple.parser.JSONParser;
import org.json.simple.parser.ParseException;

public class JsonReading {
    public static void main(String[] args){
        JsonReading jsonReading = new JsonReading();
        jsonReading.readJson("C:/testJSON.json");
    }
    public void readJson(String inFileName) {
        JSONParser parser = new JSONParser();
        try {
            Object obj = parser.parse(new FileReader(inFileName));
            JSONObject jsonObject = (JSONObject) obj;

            String name = (String) jsonObject.get("book");
            System.out.println(name);

            String author = (String) jsonObject.get("author");
            System.out.println(author);

            JSONArray reviews = (JSONArray) jsonObject.get("messages");
            Iterator<String> iterator = reviews.iterator();
            while (iterator.hasNext()) {
                System.out.println(iterator.next());
            }
        } catch (FileNotFoundException e) {
```

```
        //Your exception handling here
    } catch (IOException e) {
        //Your exception handling here
    } catch (ParseException e) {
        //Your exception handling here
    }
  }
}
```

上面的代码正常运行之后，你会在标准输出中看到 JSON 文件的内容。

1.13 使用 JSoup 从一个 URL 提取 Web 数据

当今，大量数据存在于 Web 上。在这些数据中，有些是带有结构的，有些是半结构化的，有些是不带结构的。因此，需要不同的技术来从 Web 上提取它们。有很多方法可以用来提取 Web 数据。其中，最简单易用的一种是使用一个名称为 JSoup 的外部 Java 库。本部分，我们将学习使用 JSoup 提供的一些方法来提取 Web 数据。

准备工作

开始之前，先做如下准备工作。

1. 下载 `jsoup-1.9.2.jar` 文件，并把下载到的 JAR 文件作为外部库添加到你的 Eclipse 项目中。

2. 如果你喜欢使用 Maven，请根据下载页面中的操作提示，把 JAR 文件添加到你的 Eclipse 项目中。

操作步骤

1. 创建一个名称为 `extractDataWithJsoup(String url)` 的方法。调用该方法时，你需要提供给它一个 Web 地址作为参数。这样，我们就可以从这个 URL 提取 Web 数据了。

   ```
   public void extractDataWithJsoup(String href){
   ```

2. 调用 JSoup 的 `connect()` 方法，并把要连接的 URL 地址（从该地址提取数据）传递给它。紧随其后，接连调用其他几个方法。首先是 `timeout()` 方法，带有的参数为毫秒，接下来的两个方法分别定义了连接期间的用户代理名称，以及指定是否忽略连接错误。紧接着这两个方法调用的是 get()方法，它返回一个 Document 对象。因此，我们需要事先声明一个 Document 类型的 doc 变量来保存 get()方法所返回的 Document 对象。

```
doc =
  Jsoup.connect(href).timeout(10*1000).userAgent
    ("Mozilla").ignoreHttpErrors(true).get();
```

3. 由于这段代码可能会抛出 IOException 异常，所以我们要把它放入一个 try...catch 语句块之中，如下所示：

```
Document doc = null;
try {
 doc = Jsoup.connect(href).timeout(10*1000).userAgent
    ("Mozilla").ignoreHttpErrors(true).get();
  } catch (IOException e) {
      //你的异常处理代码
}
```

由于我们不习惯于使用毫秒来计算时间，因此在代码中需要把毫秒作为时间单位时，一个更好的做法是把时间写成 10×1 000 的形式，即 10 秒。这样做，可以进一步提高代码的可读性。

4. Document 对象提供了大量用于提取数据的方法。如果你想提取 URL 的标题，你可以使用 title()方法，如下所示：

```
if(doc != null){
  String title = doc.title();
```

5. 如果你只想从 Web 页面中提取正文文本的内容，你可以接连调用 Document 对象的 body()与 text()方法，如下所示：

```
String text = doc.body().text();
```

6. 如果你想从 URL 中提取所有超链接，你可以使用 Document 对象的 select()方法，并把 a[href]作为参数提供给它。这样，你就可以一次性获取所有链接。

```
Elements links = doc.select("a[href]");
```

7. 或许你想分别处理 Web 页面中的链接，这也很容易办到。你只需要遍历所有链接，获取各个链接即可。

```
for (Element link : links) {
```

```
            String linkHref = link.attr("href");
            String linkText = link.text();
            String linkOuterHtml = link.outerHtml();
            String linkInnerHtml = link.html();
        System.out.println(linkHref + "t" + linkText + "t" +
          linkOuterHtml + "t" + linkInnterHtml);
        }
```

8. 最后，结束 if 条件语句，并关闭方法。

```
        }
    }
```

完整代码整理如下，包括完整的方法代码、相应类，以及驱动方法。

```
import java.io.IOException;
import org.jsoup.Jsoup;
import org.jsoup.nodes.Document;
import org.jsoup.nodes.Element;
import org.jsoup.select.Elements;

public class JsoupTesting {
   public static void main(String[] args){
      JsoupTesting test = new JsoupTesting();
      test.extractDataWithJsoup("Website address preceded by http://");
   }

   public void extractDataWithJsoup(String href){
      Document doc = null;
      try {
         doc = Jsoup.connect(href).timeout(10*1000).userAgent
            ("Mozilla").ignoreHttpErrors(true).get();
      } catch (IOException e) {
         //Your exception handling here
      }
      if(doc != null){
         String title = doc.title();
         String text = doc.body().text();
         Elements links = doc.select("a[href]");
         for (Element link : links) {
            String linkHref = link.attr("href");
            String linkText = link.text();
            String linkOuterHtml = link.outerHtml();
            String linkInnerHtml = link.html();
            System.out.println(linkHref + "t" + linkText + "t" +
```

```
                linkOuterHtml + "t" + linkInnterHtml);
        }
      }
    }
}
```

1.14 使用 Selenium Webdriver 从网站提取 Web 数据

Selenium 是一个基于 Java 的工具,用来进行自动软件测试或质量控制。有趣的是,Selenium 也可以用来自动提取与利用 Web 数据。本部分,我们将学习如何使用它。

准备工作

开始之前,先做如下准备。

1. 下载 `selenium-server-standalone-2.53.1.jar` 与 `seleniumjava-2.53.1.zip`。从后者提取 `selenium-java-2.53.1.jar` 文件。把这两个 JAR 文件作为外部 Java 库添加到你的 Eclipse 项目中。

2. 下载并安装 Firefox 47.0.1。注意选择与自身操作系统相对应的版本。

 由于 Selenium 与 Firefox 之间存在版本冲突问题,当你使用某个特定版本运行代码时,请记得把 Firefox 的自动下载更新与安装选项关闭。

操作步骤

1. 创建一个名称为 `extractDataWithSelenium(String)` 的方法,它带有一个 `String` 类型的参数,用来指定等待提取数据的 URL 地址。我们可以从 URL 提取的数据类型多种多样,比如标题、头部,以及下拉选择框中的值。本部分,我们将只讲解如何提取 Web 页面中的文本部分。

   ```
   public String extractDataWithSelenium(String url){
   ```

2. 接着,使用如下代码创建一个 Firefox Web 驱动器。

   ```
   WebDriver driver = new FirefoxDriver();
   ```

3. 调用 WebDriver 对象的 `get()` 方法,并把 URL 地址传递给它。

   ```
   driver.get("http://cogenglab.csd.uwo.ca/rushdi.htm");
   ```

4. 我们可以使用 xpath 查看网页文本，其中 id 的值就是内容，如图 1-2 和图 1-3 所示。

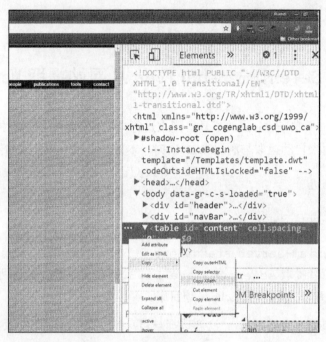

图 1-2

图 1-3

5. 调用 findElement() 方法找到这个特定元素，并且返回一个 WebElement 对象。创建一个名称为 webElement 的 WebElement 对象来保存这个返回值。

```
WebElement webElement = driver.findElement(By.xpath("//*
```

```
            [@id='content']"));
```

6. WebElement 对象有一个名为 getText() 的方法。调用这个方法，获取 Web 页面中的文本，并把文本放入一个 String 类型的变量中，如下所示：

```
String text = (webElement.getText());
```

7. 最后，返回 String 变量，关闭方法。

```
    }
```

完整代码整理如下，包括完整的方法代码、相应类，以及 main() 驱动方法。

```
import org.openqa.selenium.By;
import org.openqa.selenium.WebDriver;
import org.openqa.selenium.WebElement;
import org.openqa.selenium.firefox.FirefoxDriver;

public class TestSelenium {
   public String extractDataWithSelenium(String url) {
      WebDriver driver = new FirefoxDriver();
      driver.get("http://cogenglab.csd.uwo.ca/rushdi.htm");
      WebElement webElement = driver.findElement(By.xpath("//*
         [@id='content']"));
      System.out.println(webElement.getText());
      return url;
   }

   public static void main(String[] args){
      TestSelenium test = new TestSelenium();
      String webData = test.extractDataWithSelenium
         ("http://cogenglab.csd.uwo.ca/rushdi.htm");
      //process webData
   }
}
```

Selenium 与 Firefox 之间存在兼容性的问题。有些版本的 Selenium 无法与某些版本的 Firefox 协同工作。上面讲解中所使用的 Selenium 与 Firefox 版本能够很好地协同工作，但是这并不能保证示例代码在其他版本的 Selenium 与 Firefox 上也能正常工作。

由于 Selenium 与 Firefox 之间存在版本冲突问题，当你使用某个特定版本运行代码时，请记得把 Firefox 的自动下载更新与安装选项关闭。

1.15 从 MySQL 数据库读取表格数据

数据也可能存储在数据库的表格中。本部分，我们将演示如何从 MySQL 的数据表中读取数据。

准备工作

开始之前，先做如下准备。

1. 下载并安装 MySQL community server。本部分使用的是 MySQL 5.7.15 版本。
2. 创建一个名称为 `data_science` 的数据库。在这个数据库中，创建一个名称为 books 的数据表，包含如图 1-4 所示的数据。

id	book_name	author_name	date_created
1	The Hunger Games	Suzanne Collins	2008-09-14 00:00:00
2	Harry Potter and the Sorcerer's Stone	J.K. Rowling	1997-07-30 00:00:00
3	Divergent	Veronica Roth	2011-04-25 00:00:00

图 1-4

本部分中字段类型是什么并不重要，但是字段名称必须与上表给出的那些完全一样。

3. 下载独立于平台的 MySQL JAR 文件，并将其作为外部库添加到你的 Java 项目中。这里使用的是 5.1.39 版本。

操作步骤

1. 创建一个名为 `readTable(String user, String password, String server)` 的方法，带有的 3 个参数分别是用户名、密码，以及你的 MySQL 数据库的服务器名称。

    ```
    public void readTable(String user, String password, String
        server){
    ```

2. 创建一个 MySQL 数据源，并且通过这个数据源，设置用户名、密码与服务器名。

    ```
    MysqlDataSource dataSource = new MysqlDataSource();
    dataSource.setUser(user);
    dataSource.setPassword(password);
    ```

```
dataSource.setServerName(server);
```

3. 在 `try` 语句块中，创建数据库连接。通过这个连接，创建一条语句，用来执行 SELECT 查询，以便从数据表中获取信息。返回的查询结果将存储在结果集中。

```
try{
  Connection conn = dataSource.getConnection();
  Statement stmt = conn.createStatement();
  ResultSet rs = stmt.executeQuery("SELECT * FROM
  data_science.books");
```

4. 接着，遍历结果集，通过指定列名提取每个列。请注意，在使用相应方法提取列值之前，需要事先知道每个列字段的数据类型。比如，由于我们知道 ID 字段的数据类型是整型，所以我们可以选用 `getInt()` 方法来获取它。

```
while (rs.next()){
  int id = rs.getInt("id");
  String book = rs.getString("book_name");
  String author = rs.getString("author_name");
  Date dateCreated = rs.getDate("date_created");
  System.out.format("%s, %s, %s, %sn", id, book, author,
    dateCreated);
}
```

5. 遍历结束后，关闭结果集、查询语句与连接。

```
rs.close();
  stmt.close();
  conn.close();
```

6. 在从数据表读取数据过程中可能会出现异常，使用 **catch** 语句对可能出现的异常进行处理。最后关闭方法。

```
}catch (Exception e){
    // 你的异常处理代码
  }
}
```

完整代码整理如下，包括完整的方法代码、相应类，以及驱动方法。

```
import java.sql.*;
```

```java
import com.mysql.jdbc.jdbc2.optional.MysqlDataSource;
public class TestDB{
    public static void main(String[] args){
        TestDB test = new TestDB();
        test.readTable("your user name", "your password", "your MySQL
            server name");
    }
    public void readTable(String user, String password, String server)
        {
        MysqlDataSource dataSource = new MysqlDataSource();
        dataSource.setUser(user);
        dataSource.setPassword(password);
        dataSource.setServerName(server);
        try{
            Connection conn = dataSource.getConnection();
            Statement stmt = conn.createStatement();
            ResultSet rs = stmt.executeQuery("SELECT * FROM
                data_science.books");
            while (rs.next()){
                int id = rs.getInt("id");
                String book = rs.getString("book_name");
                String author = rs.getString("author_name");
                Date dateCreated = rs.getDate("date_created");
                System.out.format("%s, %s, %s, %sn", id, book,
                    author, dateCreated);
            }
            rs.close();
            stmt.close();
            conn.close();
        }catch (Exception e){
            //这里是你的异常处理代码
        }
    }
}
```

上面的代码正常执行之后，将会把数据表中的数据显示出来。

第 2 章
为数据建立索引与搜索数据

本章涵盖如下内容：

- 使用 Apache Lucene 为数据建立索引；
- 使用 Apache Lucene 搜索带索引的数据。

2.1 简介

本章，我们将学习两部分非常重要的内容。第一部分学习如何为数据建立索引，第二部分学习如何搜索带索引的数据，这部分内容与第一部分紧密相连。

在为数据建立索引以及搜索数据的过程中，我们将使用 Apache Lucene。它是一个免费、开源的 Java 软件库，主要用来进行信息检索。Apache Lucene 由 Apache 软件基金会提供支持与发布，遵守 Apache 软件许可证。

许多现代搜索平台与爬虫工具在后端都使用 Apache Lucene 为数据建立索引以及对数据进行搜索，比如 Apache Solr、ElasticSearch、Apache Nutch。因此，任何想学习这些搜索平台的数据科学家都能从本章的这两部分内容中获益。

2.2 使用 Apache Lucene 为数据建立索引

本部分，我们将演示如何使用 Apache Lucene 为大量数据建立索引。若要实现快速搜索数据，第一步就是为这些数据建立索引。实际上，Lucene 使用的是倒排全文索引。也就是说，Lucene 会考察所有文档，把它们拆分成单词或标记，然后为每一个标记建立索引，这样在搜索某个词语时，它能事先准确地知道要查找哪一个文档。

准备工作

开始之前，先做如下准备。

1. 进入 Lucene 的下载页面，单击 Download 按钮，下载 Apache Lucene。写作本书之时，Lucene 最新版本为 6.4.1。如图 2-1 所示，在单击 Download 按钮之后，你将转到它的镜像网站。

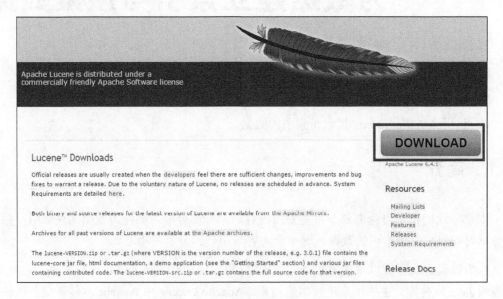

图 2-1

2. 从中选择一个合适的镜像进行下载。单击镜像网站，你将跳转到发布目录下。如图 2-2 所示，选择 `lucene-6.4.1.zip` 文件，将其下载到你的电脑中。

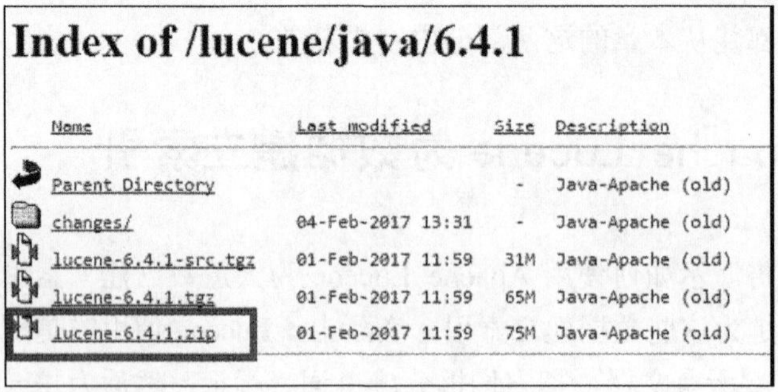

图 2-2

3. 下载完成后，进行加压缩。你将看到一个组织很好的文件夹，该文件夹包含的内容

2.2 使用 Apache Lucene 为数据建立索引 37

如图 2-3 所示。

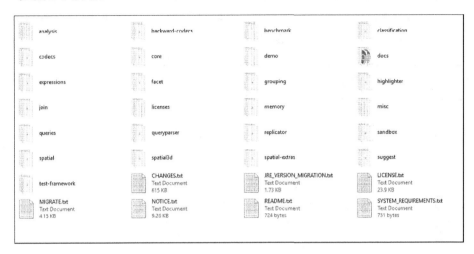

图 2-3

4. 打开 Eclipse，创建一个名称为 LuceneTutorial 的项目。为此，打开 Eclipse，在 File 中依次选择"**New...|Java Project**"，填入项目名称，单击 **Finish**，完成项目创建（见图 2-4）。

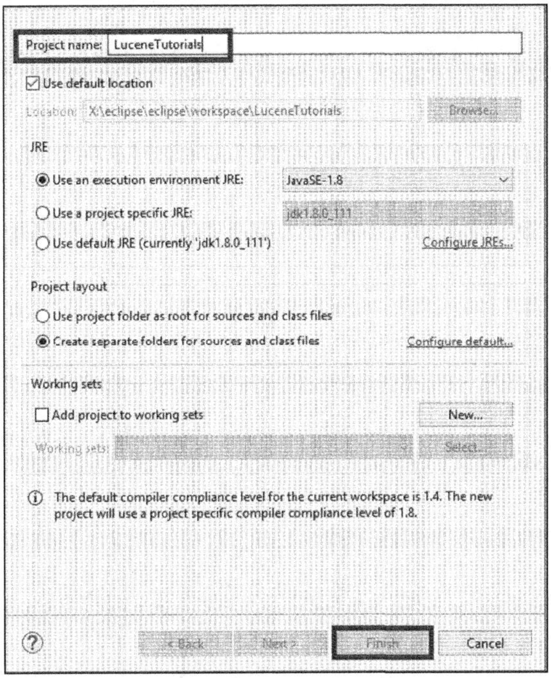

图 2-4

5. 接下来，把本部分需要的 JAR 文件作为外部库添加到你的项目中。如图 2-5 所示，在 **Package Explorer** 中，使用右键单击项目名称，依次选择 **Build Path| Configure Build Path...**，打开项目属性窗口。

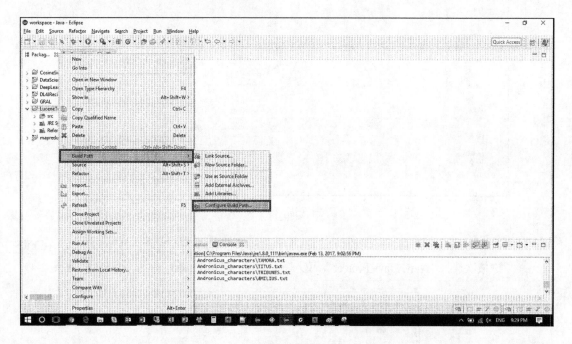

图 2-5

6. 如图 2-6 所示，单击 "Add External Jars" 按钮，然后从 Lucene 6.4.1 中添加如下 JAR 文件。

    - `lucene-core-6.4.1.jar`：位于 `lucene-6.4.1\core` 目录之下。

    - `lucene-queryparser-6.4.1.jar`：位于 `lucene-6.4.1\queryparser` 目录之下。

    - `Lucene-analyzers-common-6.4.1.jar`：位于 `lucene-6.4.1\analysis\common` 目录之下。

添加完 JAR 文件之后，单击 OK 按钮。

7. 在建立索引时，我们会用到威廉·莎士比亚的著作。打开浏览器，前往 `http://norvig.com/ngrams/`，你将打开一个名为 Natural Language Corpus Data: Beautiful Data 的页面。在 Files for Download 版块中，你会看到一个名为 shakespeare 的.txt 文件，将其下载到你的电脑中。

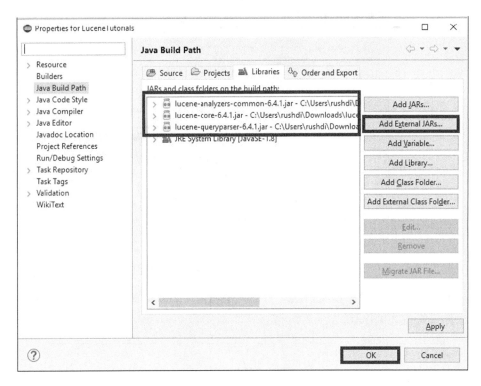

图 2-6

8. 如图 2-7 所示，下载完成后，进行解压缩，里面包含 3 个文件夹，分别是 comedies、historical、tragedies。

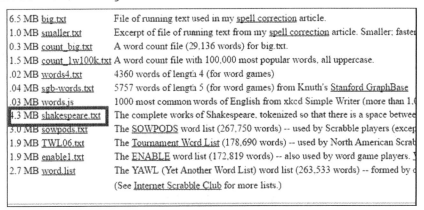

图 2-7

9. 在你的项目目录中，新建一个文件夹。如图 2-8 所示，在 Eclipse 中，右键单击项目，依次选择 **New|Folder**，在文件夹名称中，输入 input，单击 **Finish** 按钮。

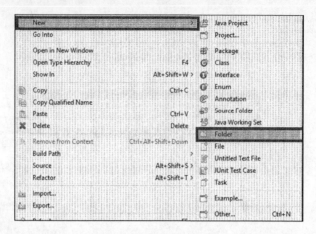

图 2-8

10. 把步骤 8 中的 `shakespeare.txt` 复制到第 9 步创建的文件夹中。

11. 根据第 9 步的操作步骤,创建另外一个名为 `index` 的文件夹。此时,你的项目文件夹如图 2-9 所示。

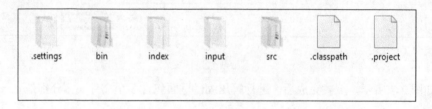

图 2-9

至此,所有准备工作就完成了。

操作步骤

1. 如图 2-10 所示,在项目中创建一个包,命名为 `org.apache.lucene.demo`,然后在这个包中创建一个 Java 文件,命名为 `IndexFiles.java`。

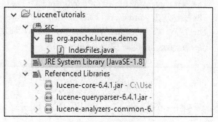

图 2-10

2. 在刚创建的 Java 文件中，新建一个名为 IndexFiles 的类。

   ```
   public class IndexFiles {
   ```

3. 接下来要编写的第一个方法是 indexDocs。该方法将使用指定的索引写者为给定的任意文件建立索引。如果为这个方法提供的参数是一个目录，这个方法就会遍历给定目录下的所有文件与子目录。这个方法会为每个输入文件建立索引。

   > 相对而言，这个方法的执行速度比较慢，因此你要把多个文档放入你的输入文件中，以便获得更好的执行性能。

   ```
   static void indexDocs(final IndexWriter writer, Path path)
       throws IOException {
   ```

   - writer 是索引写者，用来编写索引，其中存储着给定文件或目录的信息
   - path 是待建索引的文件，或目录（包含着待建索引的文件）。

4. 如果提供的是目录，这个目录会被递归遍历。

   ```
   if (Files.isDirectory(path)) {
     Files.walkFileTree(path, new SimpleFileVisitor<Path>() {
   ```

5. 接下来，重载名为 visitFile 的方法，用来根据给定的路径与基本的文件属性访问文件或目录。

   ```
   @Override
     public FileVisitResult visitFile(Path file,
       BasicFileAttributes attrs) throws IOException {
   ```

6. 然后，调用一个名为 indexDoc 的静态方法，稍后我们会创建它。这里我们有意把 catch 语句块留空，当无法为文件建立索引时应该做什么，请你自己决定。

   ```
   try {
       indexDoc(writer, file,
         attrs.lastModifiedTime().toMillis());
     } catch (IOException ignore) {
   }
   ```

7. 从 `visitFile` 方法返回。

   ```
   return FileVisitResult.CONTINUE;
   }
   ```

8. 关闭语句块。

   ```
   }
     );
   }
   ```

9. 在 `else` 语句块中，调用 `indexDoc` 方法。请注意，在 `else` 语句块中，我们处理的是文件，而不是目录。

   ```
   else {
     indexDoc(writer, path,
       Files.getLastModifiedTime(path).toMillis());
   }
   ```

10. 关闭 `indexDocs()` 方法。

    ```
    }
    ```

11. 接下来，创建为单个文档建立索引的方法。

    ```
    static void indexDoc(IndexWriter writer, Path file, long
      lastModified) throws IOException {
    ```

12. 首先，在 `try` 语句块中新建一个空文档。

    ```
    try (InputStream stream = Files.newInputStream(file)) {
      Document doc = new Document();
    ```

13. 接着，添加文件路径字段，输入"path"作为字段名。这个字段是可检索或可索引的。然而，请注意，我们没有对字段做标记，也没有为词频或位置信息建立索引。

    ```
    Field pathField = new StringField("path", file.toString(),
      Field.Store.YES);
    doc.add(pathField);
    ```

14. 添加文件的最后修改日期，字段名是"modified"。

```
doc.add(new LongPoint("modified", lastModified));
```

15. 添加文件内容到名为 `contents` 的字段中。你指定的"读者"（reader）将保证文件的文本被标记化与索引化，但未被保存。

    ```
    doc.add(new TextField("contents", new BufferedReader(new
       InputStreamReader(stream, StandardCharsets.UTF_8))));
    ```

     如果文件采用的不是 UTF-8 编码，那么搜索特殊字符时会失败。

16. 为文件创建索引。

    ```
    if (writer.getConfig().getOpenMode() == OpenMode.CREATE) {
        System.out.println("adding " + file);
        writer.addDocument(doc);
    }
    ```

17. 有可能文档已经建立好索引了。对于这些情况，我们可以在 `else` 语句块中进行处理。如果存在，我们将使用 `updateDocument` 代替旧的那个，以匹配准确的路径。

    ```
    else {
        System.out.println("updating " + file);
        writer.updateDocument(new Term("path", file.toString()),
          doc);
    }
    ```

18. 关闭 try 语句块与方法。

    ```
        }
    }
    ```

19. 接下来，让我们创建类的 main 方法。

    ```
    public static void main(String[] args) {
    ```

20. 运行程序时，我们将通过控制台提供如下 3 个选项。

- 第一个选项是 index,该参数是包含索引的文件夹。
- 第二个选项是 docs,该参数是包含文本文件的文件夹。
- 最后一个选项是 update,这个参数表示你是想创建新索引还是更新旧索引。

为了保存这 3 个参数的值,创建并初始化如下 3 个变量。

```
String indexPath = "index";
String docsPath = null;
boolean create = true;
```

21. 设置 3 个选项的值。

```
for(int i=0;i<args.length;i++) {
 if ("-index".equals(args[i])) {
    indexPath = args[i+1];
    i++;
 } else if ("-docs".equals(args[i])) {
    docsPath = args[i+1];
    i++;
 } else if ("-update".equals(args[i])) {
    create = false;
 }
}
```

22. 设置文档目录。

```
final Path docDir = Paths.get(docsPath);
```

23. 接下来,我们开始为目录中的文件建立索引。首先,设置定时器,因为我们需要对创建索引的延迟时间进行计算。

```
Date start = new Date();
```

24. 为了建立索引,创建一个目录与分析器(在本例中,我们将使用一个最基本的标准分析器和一个索引写者配置器)。

```
try {

    Directory dir = FSDirectory.open(Paths.get(indexPath));
```

```
            Analyzer analyzer = new StandardAnalyzer();
            IndexWriterConfig iwc = new IndexWriterConfig(analyzer);
```

25. 借助配置好的索引写者，根据输入为索引创建还是更新，为索引设置打开模式。如果你选择创建新索引，打开模式将为 CREATE，否则将为 CREATE_OR_APPEND。

    ```
    if (create) {
        iwc.setOpenMode(OpenMode.CREATE);
    } else {
        iwc.setOpenMode(OpenMode.CREATE_OR_APPEND);
    }
    ```

26. 创建一个索引写者。

    ```
    IndexWriter writer = new IndexWriter(dir, iwc);
    indexDocs(writer, docDir);
    ```

27. 关闭 writer。

    ```
    writer.close();
    ```

28. 到这里，我们几乎已经写完所有代码。最后计算建立索引所耗费的时间。

    ```
    Date end = new Date();
    System.out.println(end.getTime() - start.getTime() + " total
      milliseconds");
    ```

29. 关闭 try 语句块。这里我们有意把 catch 语句块留空，这样一来你就可以在建立索引的过程中出现异常时自己决定要做什么。

    ```
    } catch (IOException e) {
    }
    ```

30. 关闭 main 方法与类。

    ```
        }
    }
    ```

31. 如图 2-11 所示，在 Eclipse 中，右键单击项目，依次选择 **Run As|Run Configurations…**。

46　第 2 章　为数据建立索引与搜索数据

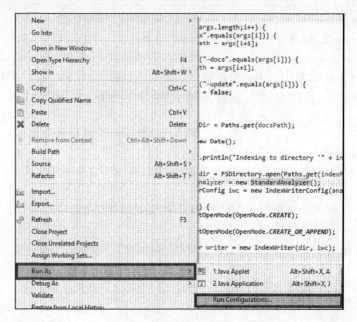

图 2-11

32. 如图 2-12 所示，在 Run Configurations 窗口中，进入 **Arguments** 选项卡。在 Program Arguments 中输入 `-docs input\ -index index\`，单击 **Run** 按钮。

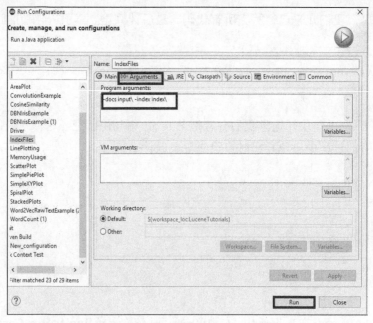

图 2-12

33. 代码输出如图 2-13 所示。

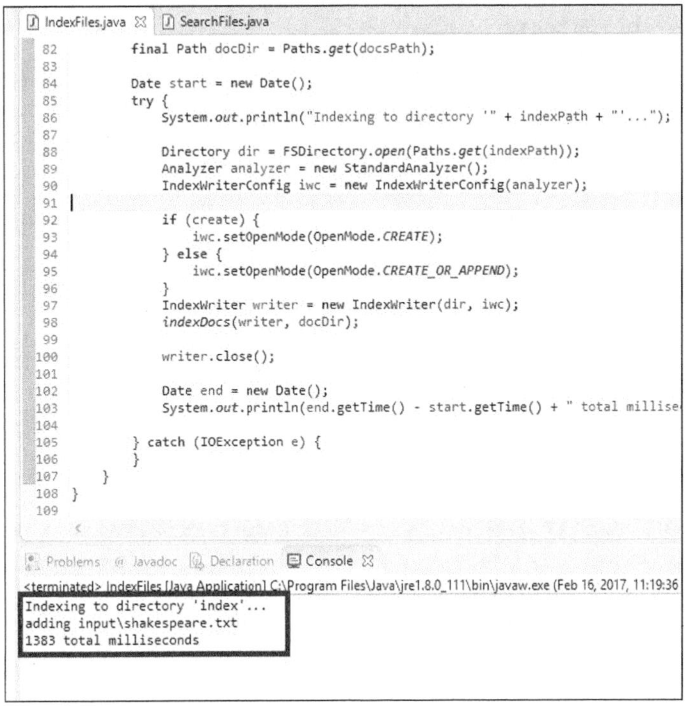

图 2-13

工作原理

完整代码整理如下：

```
package org.apache.lucene.demo;

import org.apache.lucene.analysis.Analyzer;
import org.apache.lucene.analysis.standard.StandardAnalyzer;
import org.apache.lucene.document.Document;
import org.apache.lucene.document.Field;
import org.apache.lucene.document.LongPoint;
import org.apache.lucene.document.StringField;
import org.apache.lucene.document.TextField;
import org.apache.lucene.index.IndexWriter;
import org.apache.lucene.index.IndexWriterConfig.OpenMode;
import org.apache.lucene.index.IndexWriterConfig;
import org.apache.lucene.index.Term;
import org.apache.lucene.store.Directory;
```

```java
import org.apache.lucene.store.FSDirectory;
import java.io.BufferedReader;
import java.io.IOException;
import java.io.InputStream;
import java.io.InputStreamReader;
import java.nio.charset.StandardCharsets;
import java.nio.file.FileVisitResult;
import java.nio.file.Files;
import java.nio.file.Path;
import java.nio.file.Paths;
import java.nio.file.SimpleFileVisitor;
import java.nio.file.attribute.BasicFileAttributes;
import java.util.Date;

public class IndexFiles {
   static void indexDocs(final IndexWriter writer, Path path) throws
      IOException {
     if (Files.isDirectory(path)) {
        Files.walkFileTree(path, new SimpleFileVisitor<Path>() {
           @Override
           public FileVisitResult visitFile(Path file,
              BasicFileAttributes attrs) throws IOException {
             try {
                indexDoc(writer, file,
                   attrs.lastModifiedTime().toMillis());
             } catch (IOException ignore) {
             }
             return FileVisitResult.CONTINUE;
           }
         }
         );
     } else {
        indexDoc(writer, path,
           Files.getLastModifiedTime(path).toMillis());
     }
   }

   static void indexDoc(IndexWriter writer, Path file, long
      lastModified) throws IOException {
     try (InputStream stream = Files.newInputStream(file)) {
        Document doc = new Document();
        Field pathField = new StringField("path", file.toString(),
           Field.Store.YES);
        doc.add(pathField);
        doc.add(new LongPoint("modified", lastModified));
        doc.add(new TextField("contents", new BufferedReader(new
```

```java
                InputStreamReader(stream, StandardCharsets.UTF_8))));

        if (writer.getConfig().getOpenMode() == OpenMode.CREATE) {
            System.out.println("adding " + file);
            writer.addDocument(doc);
        } else {
            System.out.println("updating " + file);
            writer.updateDocument(new Term("path", file.toString()),
               doc);
        }
    }
}
public static void main(String[] args) {
    String indexPath = "index";
    String docsPath = null;
    boolean create = true;
    for(int i=0;i<args.length;i++) {
        if ("-index".equals(args[i])) {
            indexPath = args[i+1];
            i++;
        } else if ("-docs".equals(args[i])) {
            docsPath = args[i+1];
            i++;
        } else if ("-update".equals(args[i])) {
            create = false;
        }
    }

    final Path docDir = Paths.get(docsPath);

    Date start = new Date();
    try {
        System.out.println("Indexing to directory '" + indexPath +
          "'...");

        Directory dir = FSDirectory.open(Paths.get(indexPath));
        Analyzer analyzer = new StandardAnalyzer();
        IndexWriterConfig iwc = new IndexWriterConfig(analyzer);

        if (create) {
            iwc.setOpenMode(OpenMode.CREATE);
        } else {
            iwc.setOpenMode(OpenMode.CREATE_OR_APPEND);
        }
        IndexWriter writer = new IndexWriter(dir, iwc);
        indexDocs(writer, docDir);
```

```
            writer.close();

            Date end = new Date();
            System.out.println(end.getTime() - start.getTime() + " total
                milliseconds");

        } catch (IOException e) {
        }
    }
}
```

2.3 使用 Apache Lucene 搜索带索引的数据

上一节中我们已经为数据建立好了索引，接下来我们将使用 Apache Lucene 来搜索数据。本部分的搜索代码依赖于你在上一节中创建的索引，因此你只有根据上一节的操作步骤来做，代码才能成功执行。

准备工作

1. 做完上一节的所有步骤，然后进入第 11 步中创建的索引目录，确保索引目录中存在如图 2-14 所示的索引文件。

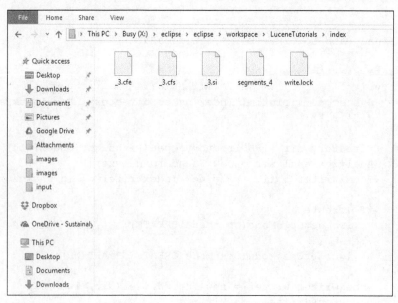

图 2-14

2. 在 `org.apache.lucene.demo` 包（该包已在上一节中创建出来）中创建一个名为 `SearchFiles` 的 Java 文件，如图 2-15 所示。

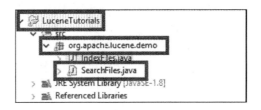

图 2-15

3. 接下来，需要在 `SearchFiles.java` 文件中添加一些代码。

操作步骤

1. 在 Eclipse 的编辑器中打开 `SearchFiles.java` 文件，创建如下类。

   ```
   public class SearchFiles {
   ```

2. 接下来，需要创建两个静态的 String 类型的变量。第一个用来保存你的 `index`（已经在上一节中创建出来）的路径，第二个用来保存待搜索的字段内容。在本例中，我们将在 `index` 的 `contents` 字段中进行搜索。

   ```
   public static final String INDEX_DIRECTORY = "index";
   public static final String FIELD_CONTENTS = "contents";
   ```

3. 开始创建 main 方法。

   ```
   public static void main(String[] args) throws Exception {
   ```

4. 通过打开 `index` 目录中的索引，创建一个 `IndexReader`。

   ```
   IndexReader reader =
     DirectoryReader.open(FSDirectory.open
       (Paths.get(INDEX_DIRECTORY)));
   ```

5. 接下来，创建一个搜索器，用来搜索索引。

   ```
   IndexSearcher indexSearcher = new IndexSearcher(reader);
   ```

6. 创建一个标准分析器。

```
Analyzer analyzer = new StandardAnalyzer();
```

7. 创建一个查询分析器,并向 `QueryParser` 构造函数提供两个参数,一个是要搜索的字段,另一个是前面创建的分析器。

    ```
    QueryParser queryParser = new QueryParser(FIELD_CONTENTS,
        analyzer);
    ```

8. 本部分,我们将使用一个预定义的搜索词。搜索中,我们将尝试查找同时包含 over-full 与 persuasion 的文档。

    ```
    String searchString = "over-full AND persuasion";
    ```

9. 使用搜索字符串,创建一个查询。

    ```
    Query query = queryParser.parse(searchString);
    ```

10. 搜索器将查看索引,看看是否能够找到搜索词。你也可以指定返回多少个搜索结果,这里我们设置为 5。

    ```
    TopDocs results = indexSearcher.search(query, 5);
    ```

11. 创建一个数组,用来保存 hits。

    ```
    ScoreDoc[] hits = results.scoreDocs;
    ```

12. 请注意,在建立索引期间,我们只用了一个文档,即 `shakespeare.txt`。所以本例中数组的长度是最大值 1。

13. 你可能也想知道搜索命中的文档数量。

    ```
    int numTotalHits = results.totalHits;
    System.out.println(numTotalHits + " total matching documents");
    ```

14. 最后,遍历 hits,获取搜索命中的文档 ID。然后,通过文档 ID,创建文档,并打印文档的路径与 Lucene 为文档计算的分数(针对你所使用的搜索词)。

    ```
    for(int i=0;i<hits.length;++i) {
     int docId = hits[i].doc;
     Document d = indexSearcher.doc(docId);
     System.out.println((i + 1) + ". " + d.get("path") + " score="
    ```

```
            + hits[i].score);
    }
```

15. 关闭方法与类

```
    }
}
```

16. 运行代码，你将看到如图 2-16 所示的输出。

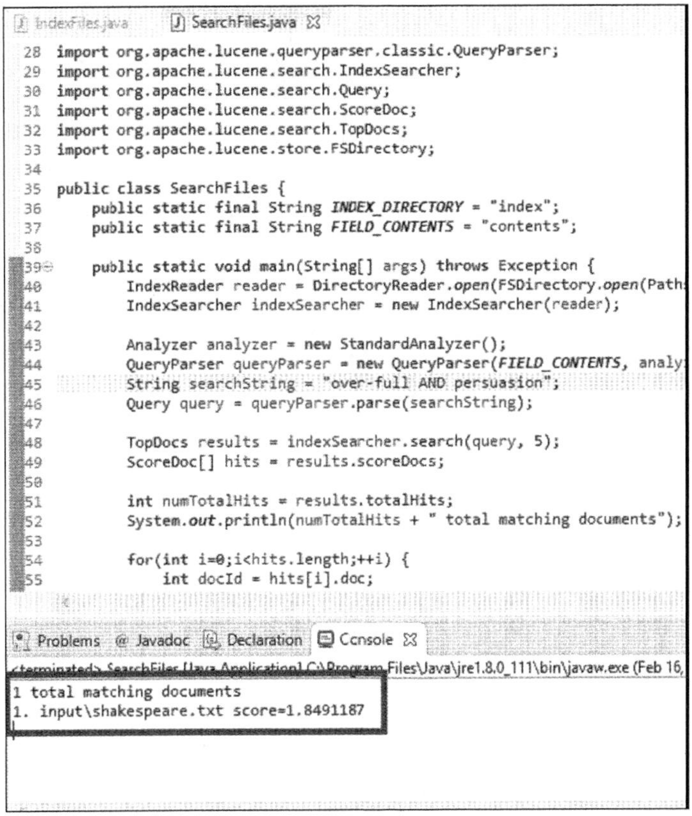

图 2-16

17. 在项目文件夹的 `input` 文件夹中，打开 `shakespeare.txt` 文件。手工进行搜索，你将看到 `over-full` 与 `persuasion` 同时出现在文档中。

18. 修改第 8 步中的 `searchString`，如下所示。

```
String searchString = "shakespeare";
```

19. 其余代码保持不变，运行代码，你将看到如图 2-17 所示的输出结果。

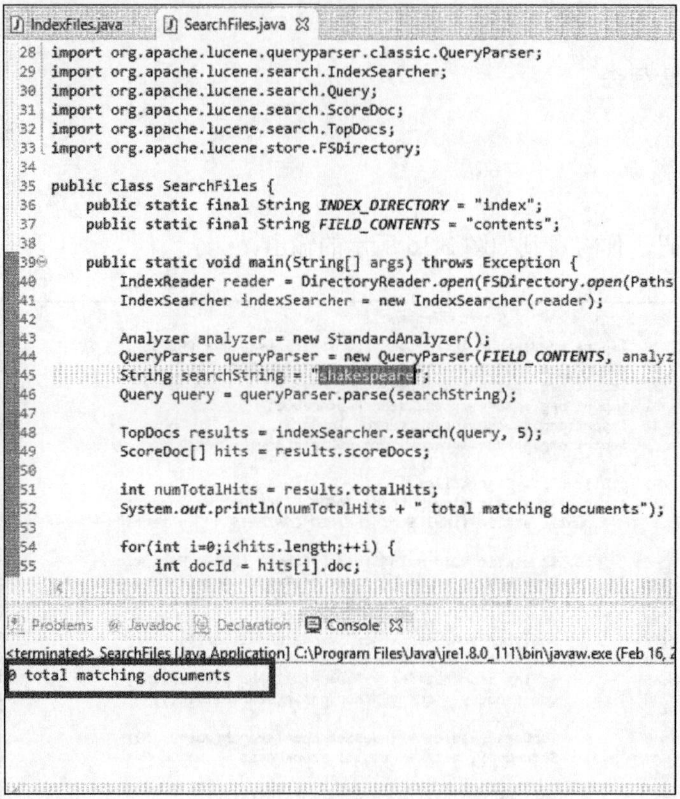

图 2-17

20. 再次打开 Shakespeare.txt 文件，检查其中是否包含 shakespeare 一词，你会发现文档中没有这个词。

完整代码整理如下：

```
package org.apache.lucene.demo;
import java.nio.file.Paths;
import org.apache.lucene.analysis.Analyzer;
import org.apache.lucene.analysis.standard.StandardAnalyzer;
import org.apache.lucene.document.Document;
import org.apache.lucene.index.DirectoryReader;
import org.apache.lucene.index.IndexReader;
import org.apache.lucene.queryparser.classic.QueryParser;
import org.apache.lucene.search.IndexSearcher;
import org.apache.lucene.search.Query;
import org.apache.lucene.search.ScoreDoc;
```

```java
import org.apache.lucene.search.TopDocs;
import org.apache.lucene.store.FSDirectory;

public class SearchFiles {
    public static final String INDEX_DIRECTORY = "index";
    public static final String FIELD_CONTENTS = "contents";

    public static void main(String[] args) throws Exception {
        IndexReader reader = DirectoryReader.open(FSDirectory.open
            (Paths.get(INDEX_DIRECTORY)));
        IndexSearcher indexSearcher = new IndexSearcher(reader);

        Analyzer analyzer = new StandardAnalyzer();
        QueryParser queryParser = new QueryParser(FIELD_CONTENTS,
            analyzer);
        String searchString = "shakespeare";
        Query query = queryParser.parse(searchString);

        TopDocs results = indexSearcher.search(query, 5);
        ScoreDoc[] hits = results.scoreDocs;

        int numTotalHits = results.totalHits;
        System.out.println(numTotalHits + " total matching documents");

        for(int i=0;i<hits.length;++i) {
            int docId = hits[i].doc;
            Document d = indexSearcher.doc(docId);
            System.out.println((i + 1) + ". " + d.get("path") + " score="
                + hits[i].score);
        }
    }
}
```

> 关于 Apache Lucene 所支持的查询语法，你可以前往其官网进行了解学习。

第 3 章
数据统计分析

本章包含如下内容：
- 生成描述性统计；
- 生成概要统计；
- 从多种分布生成概要统计；
- 计算频率分布；
- 计算字符串中的词频；
- 使用 Java8 计算字符串中的词频；
- 计算简单回归；
- 计算普通最小二乘回归；
- 计算广义最小二乘回归；
- 计算两组数据点的协方差；
- 计算两组数据点的皮尔逊相关系数；
- 执行配对 t 检验；；
- 执行卡方检验；
- 执行单因素方差分析；
- 执行 K-S 检验。

3.1 简介

统计分析是数据科学家要进行的常规活动之一。这些分析包括（但不限于）描述性分析、频率分布、简单与多重回归、相关与协方差，以及数据分布中的统计显著性。幸运的是，Java 提供了许多库，用于支持数据统计分析，借助这些库，我们只需编写几行代码就能对数据进行统计分析。本章包含 15 个小节，讲解数据科学家如何使用 Java 对数据进行统计分析。

请注意，本章讲解的重点是使用 Java 对数据进行基本的统计分析，但是你完全可以使用 Java 进行线性代数、数值分析、特殊函数、复数计算、几何学、曲线拟合、微分方程等复杂计算。

开始讲解本章内容之前，需要先做如下准备。

1. 下载 Apache Commons Math 3.6.1。
2. 如果你想使用旧版本，请前往官网下载，如图 3-1 所示。

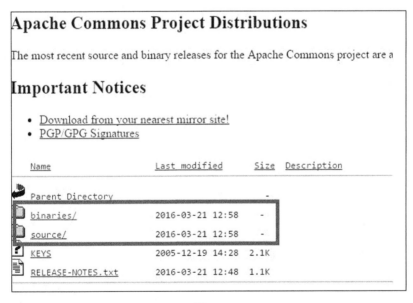

图 3-1

3. 如图 3-2 所示，下载完成后，把 JAR 文件作为外部文件添加到你的 Eclipse 项目中。

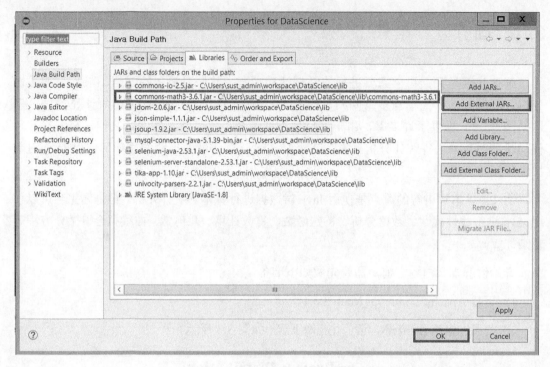

图 3-2

Apache Commons Math 3.6.1 的 stat 包的内容非常丰富，并且得到很好的优化。使用这个包能够生成如下描述性统计：

- 算术与几何平均数；
- 方差和标准差；
- 和、积、对数求和、平方和；
- 最小值、最大值、中位数与百分位数；
- 偏度和峰度；
- 一阶、二阶、三阶、四阶矩量。

而且，根据官方网站的说法，这些方法都经过优化，执行时占用的内存更少。

除了百分位数与中位数之外，所有这些统计量在计算时都不需要在内存中维护输入数据值的完整列表。

3.2 生成描述性统计

描述性统计用来概述样本，其发展通常不基于概率理论。相比之下，推论统计主要用于从一个代表性的样本来推论群体的特征。本部分，我们将学习如何使用 Java 从较少的样本来生成描述性统计。

关于描述性分析，这里我们不会谈论太多的内容，接下来的讲解主要集中在描述性统计的一个子集上。

操作步骤

1. 创建名为 `getDescStats` 的方法，其参数是 `double` 类型的数组，保存着要进行描述性统计的数值。

    ```
    public void getDescStats(double[] values){
    ```

2. 创建一个 `DescriptiveStatistics` 类型的对象。

    ```
    DescriptiveStatistics stats = new DescriptiveStatistics();
    ```

3. 遍历 `double` 型数组中的所有值，并且把它们添加到 `DescriptiveStatistic` 对象。

    ```
    for( int i = 0; i < values.length; i++) {
        stats.addValue(values[i]);
    }
    ```

4. 在 Apache Commons Math 库的 `DescriptiveStatistics` 类中有许多方法用来计算一组数值的平均值、标准差与中位数。只要调用这些方法，就能获取数据的描述性统计。最后，关闭方法。

    ```
    double mean = stats.getMean();
    double std = stats.getStandardDeviation();
    double median = stats.getPercentile(50);
    System.out.println(mean + "\t" + std + "\t" + median);
    }
    ```

完整代码整理如下，包含相应类与驱动方法。

```
import org.apache.commons.math3.stat.descriptive.DescriptiveStatistics;

public class DescriptiveStats {
    public static void main(String[] args){
        double[] values = {32, 39, 14, 98, 45, 44, 45, 34, 89, 67, 0,
            15, 0, 56, 88};
        DescriptiveStats descStatTest = new DescriptiveStats();
        descStatTest.getDescStats(values);
    }
    public void getDescStats(double[] values){
        DescriptiveStatistics stats = new DescriptiveStatistics();
        for( int i = 0; i < values.length; i++) {
            stats.addValue(values[i]);
        }
        double mean = stats.getMean();
        double std = stats.getStandardDeviation();
        double median = stats.getPercentile(50);
        System.out.println(mean + "\t" + std + "\t" + median);
    }
}
```

在进行描述性统计分析时，为了保证线程安全，我们可以创建一个 SynchronizedDescriptiveStatistics 实例，如下：DescriptiveStatistics stats = new SynchronizedDescriptiveStatistics();

3.3 生成概要统计

借助 SummaryStatistics 类，我们可以对数据进行概要统计。这个类与上一节中使用的 DescriptiveStatistics 类相似，但与 DescriptiveStatistics 类最大的不同是，SummaryStatistics 类不把数据存在内存中。

操作步骤

1. 类似于上一节，创建一个名为 getSummaryStats 的方法，其参数是 double 类型的数组。

    ```
    public void getSummaryStats(double[] values){
    ```

2. 创建一个 SummaryStatistics 类的对象。

```
        SummaryStatistics stats = new SummaryStatistics();
```

3. 把所有值添加到 SummaryStatistics 类的对象中。

```
        for( int i = 0; i < values.length; i++) {
            stats.addValue(values[i]);
        }
```

4. 最后，调用 SummaryStatistics 类中的方法，为这些值生成概要统计。在用完这些统计量之后，关闭该方法。

```
        double mean = stats.getMean();
        double std = stats.getStandardDeviation();
        System.out.println(mean + "\t" + std);
        }
```

完整代码整理如下，包含相应类与驱动方法。

```
import org.apache.commons.math3.stat.descriptive.SummaryStatistics;

public class SummaryStats {
    public static void main(String[] args){
        double[] values = {32, 39, 14, 98, 45, 44, 45, 34, 89, 67, 0, 15,
          0, 56, 88};
        SummaryStats summaryStatTest = new SummaryStats();
        summaryStatTest.getSummaryStats(values);
    }
    public void getSummaryStats(double[] values){
        SummaryStatistics stats = new SummaryStatistics();
        for( int i = 0; i < values.length; i++) {
              stats.addValue(values[i]);
        }
        double mean = stats.getMean();
        double std = stats.getStandardDeviation();
        System.out.println(mean + "\t" + std);
    }
}
```

3.4 从多种分布生成概要统计

本部分，我们将创建一个 AggregateSummaryStatistics 实例，用来为样本数据进行全面统计以及概要统计（SummaryStatistics）。

操作步骤

1. 创建一个名为 getAggregateStats 的方法，它拥有的两个参数，都是 double 类型数组。这两个数组包含两组不同的数据。

   ```
   public void getAggregateStats(double[] values1, double[]
     values2){
   ```

2. 创建一个 AggregateSummaryStatistics 类型的对象。

   ```
   AggregateSummaryStatistics aggregate = new
   AggregateSummaryStatistics();
   ```

3. 为了从两种分布生成概要统计，创建两个 SummaryStatistics 类的对象。

   ```
   SummaryStatistics firstSet =
     aggregate.createContributingStatistics();
   SummaryStatistics secondSet =
     aggregate.createContributingStatistics();
   ```

4. 把两种分布的值添加到上面创建的两个对象中。

   ```
   for(int i = 0; i < values1.length; i++) {
      firstSet.addValue(values1[i]);
   }
   for(int i = 0; i < values2.length; i++) {
      secondSet.addValue(values2[i]);
   }
   ```

5. 使用 AggregateSummaryStatistics 类的方法，从两种分布生成汇总统计。最后，在使用完统计量之后，关闭方法。

   ```
   double sampleSum = aggregate.getSum();
   double sampleMean = aggregate.getMean();
   double sampleStd= aggregate.getStandardDeviation();
   System.out.println(sampleSum + "\t" + sampleMean + "\t" +
     sampleStd);
   }
   ```

完整代码整理如下：

```java
import org.apache.commons.math3.stat.descriptive.
  AggregateSummaryStatistics;
import org.apache.commons.math3.stat.descriptive.SummaryStatistics;

public class AggregateStats {
  public static void main(String[] args){
    double[] values1 = {32, 39, 14, 98, 45, 44, 45};
    double[] values2 = {34, 89, 67, 0, 15, 0, 56, 88};
    AggregateStats aggStatTest = new AggregateStats();
    aggStatTest.getAggregateStats(values1, values2);
  }
  public void getAggregateStats(double[] values1, double[] values2){
    AggregateSummaryStatistics aggregate = new
    AggregateSummaryStatistics();
    SummaryStatistics firstSet =
      aggregate.createContributingStatistics();
    SummaryStatistics secondSet =
      aggregate.createContributingStatistics();
    for(int i = 0; i < values1.length; i++) {
       firstSet.addValue(values1[i]);
    }
    for(int i = 0; i < values2.length; i++) {
       secondSet.addValue(values2[i]);
    }
    double sampleSum = aggregate.getSum();
    double sampleMean = aggregate.getMean();
    double sampleStd= aggregate.getStandardDeviation();
    System.out.println(sampleSum + "\t" + sampleMean + "\t" +
      sampleStd);
    }
  }
```

更多内容

本部分介绍的方法有一些缺点，如下所示：

- 每次调用 addValue() 方法时，调用必须与 aggregate 所维护的 SummaryStatistics 实例保持同步；
- 每次添加一个值，都会更新 aggregate 与样本。

为了克服这些缺点，你可以在这个类中使用 `static aggregate` 方法。

3.5 计算频率分布

`Frequency` 类中有多个方法可以用来计算桶中的数据实例数目，以及数据实例的唯一编号等。`Frequency` 的接口非常简单，大部分情况下，只需要使用少数几行代码就能执行指定的计算。

并且，支持字符串、整型、长整型、字符这些值类型。

 累积频率的默认排序是自然顺序（Natural ordering），但是通过向构造函数提供一个 Comparator 可以改变它。

操作步骤

1. 创建一个名为 `getFreqStats` 的方法，其参数为 `double` 类型数组。我们将为这个数组中的值计算频率分布。

   ```
   public void getFreqStats(double[] values){
   ```

2. 创建一个 `Frequency` 类的对象。

   ```
   Frequency freq = new Frequency();
   ```

3. 把 `double` 数组的值添加到这个对象。

   ```
   for( int i = 0; i < values.length; i++) {
       freq.addValue(values[i]);
   }
   ```

4. 为数组中的每个值生成频数。

   ```
   for( int i = 0; i < values.length; i++) {
     System.out.println(freq.getCount(values[i]));
   }
   ```

5. 最后，关闭方法。

   ```
   }
   ```

完整代码整理如下：

```java
import org.apache.commons.math3.stat.Frequency;

public class FrequencyStats {
    public static void main(String[] args){
        double[] values = {32, 39, 14, 98, 45, 44, 45, 34, 89, 67, 0, 15,
            0, 56, 88};
        FrequencyStats freqTest = new FrequencyStats();
        freqTest.getFreqStats(values);
    }
    public void getFreqStats(double[] values){
        Frequency freq = new Frequency();
        for( int i = 0; i < values.length; i++) {
            freq.addValue(values[i]);
        }

        for( int i = 0; i < values.length; i++) {
            System.out.println(freq.getCount(values[i]));
        }
    }
}
```

3.6 计算字符串中的词频

本部分与本章的其他部分有着很大的不同，因为这部分中我们要处理的是字符串，计算字符串中的词频。我们可以使用 Apache Commons Math 与 Java 8 来完成这项工作。本节将使用 Apache Commons Math 这个外部库来进行，而在下一节中我们会使用 Java 8 来进行这项工作。

操作步骤

1. 创建一个名为 getFreqStats 的方法，其参数为字符串数组，包含着字符串中的所有单词。

   ```java
   public void getFreqStats(String[] words){
   ```

2. 创建一个 Frequency 类的对象。

   ```java
   Frequency freq = new Frequency();
   ```

3. 把所有单词添加到 Frequency 对象。

```
for( int i = 0; i < words.length; i++) {
   freq.addValue(words[i].trim());
}
```

4. 调用 Frequency 类的 getCount() 方法，针对每个单词计算频率。处理完成后，关闭方法。

```
for( int i = 0; i < words.length; i++) {
   System.out.println(words[i] + "=" +
      freq.getCount(words[i]));
   }
}
```

完整代码整理如下：

```
import org.apache.commons.math3.stat.Frequency;

public class WordFrequencyStatsApache {
  public static void main(String[] args){
    String str = "Horatio says 'tis but our fantasy, "
        + "And will not let belief take hold of him "
        + "Touching this dreaded sight, twice seen of us. "
        + "Therefore I have entreated him along, 35"
        + "With us to watch the minutes of this night, "
        + "That, if again this apparition come, "
        + "He may approve our eyes and speak to it.";
    String[] words = str.toLowerCase().split("\\W+");
    WordFrequencyStatsApache freqTest = new
      WordFrequencyStatsApache();
    freqTest.getFreqStats(words);

  }
  public void getFreqStats(String[] words){
  Frequency freq = new Frequency();
  for( int i = 0; i < words.length; i++) {
     freq.addValue(words[i].trim());
     }

     for( int i = 0; i < words.length; i++) {
        System.out.println(words[i] + "=" + freq.getCount(words[i]));
     }
   }
}
```

工作原理

本部分会把每个单词以及它们在字符串中出现的频率打印出来,这样在输出中单词会出现两次。当在最后一个 `for` 循环中处理频率时,你需要使用编程方法来避免重复输出的问题。

比如,在下一节中,我们会使用 `Map` 这种数据结构来解决单词重复的问题。如果单词的顺序不重要,你可以使用 `HashMap` 这个数据结构;如果单词顺序很重要,请使用 `TreeMap` 这种数据结构。

3.7 使用 Java 8 计算字符串中的词频

本部分不会使用 Apache Commons Math 库来计算给定字符串中的词频,取而代之,我们使用 Java 8 引入的核心库与机制来进行这项工作。

 在 Java 中有很多方法可以用来计算词频。建议各位了解一下在 Java 8 之前的版本中用来统计词频的实现都有哪些。

操作步骤

1. 创建一个名为 `getFreqStats` 的方法,它带有一个 `String` 类型的参数。我们将统计这个字符串中单词出现的频率。

   ```
   public void getFreqStats(String str){
   ```

2. 从给定的字符串创建一个 `Stream`。本例中,我们将把字符串转换为小写,并使用一个正则表达式\W+识别单词。把字符串转换成流的过程是并行进行的。

   ```
   Stream<String> stream =
     Stream.of(str.toLowerCase().split("\\W+")).parallel();
   ```

3. 调用 `Stream` 类的 `collect()` 方法来采集单词及其它们的频率。请注意,采集结果会被赋给一个 `Map` 对象,泛型中包含 `String` 与 `Long`,其中前者保存单词,后者保存单词的频率。

   ```
   Map<String, Long> wordFreq =
   ```

```
        stream.collect(Collectors.groupingBy
            (String::toString,Collectors.counting()));
```

4. 最后，使用 `forEach` 一次性打印出 Map 的内容，关闭方法。

```
        wordFreq.forEach((k,v)->System.out.println(k + "=" + v));
    }
```

示例的完整代码整理如下：

```
import java.util.Map;
import java.util.stream.Collectors;
import java.util.stream.Stream;
public class WordFrequencyStatsJava {

    public static void main(String[] args){
        String str = "Horatio says 'tis but our fantasy, "
            + "And will not let belief take hold of him "
            + "Touching this dreaded sight, twice seen of us. "
            + "Therefore I have entreated him along, 35"
            + "With us to watch the minutes of this night, "
            + "That, if again this apparition come, "
            + "He may approve our eyes and speak to it.";
        WordFrequencyStatsJava freqTest = new WordFrequencyStatsJava();
          freqTest.getFreqStats(str);
    }
    public void getFreqStats(String str){
        Stream<String> stream =
Stream.of(str.toLowerCase().split("\\W+")).parallel();
        Map<String, Long> wordFreq = stream
.collect(Collectors.groupingBy(String::toString,Collectors.counting()));
        wordFreq.forEach((k,v)->System.out.println(k + "=" + v));
    }
}
```

3.8 计算简单回归

`SimpleRegression` 类支持带有一个自变量的普通最小二乘回归（ordinary least squares regression）：y = intercept + slope × x，其中 `intercept` 是可选参数。这个类还能为 `intercept` 提供标准误差。你可以向模型一个个地添加观测值(x,y)对，也可以通过一个二维数组提供它们。本部分中，我们会逐个添加数据点。

由于观测值并不存储在内存中,所以对于可添加到模型的观测值的数量没有任何限制。

操作步骤

1. 为了计算简单回归,创建一个名为 calculateRegression 的方法,它带有一个 double 型的二维数组。这个数组表示一系列的(x,y)值。

    ```
    public void calculateRegression(double[][] data){
    ```

2. 创建一个 SimpleRegression 类的对象,并添加数据。

    ```
    SimpleRegression regression = new SimpleRegression();
    regression.addData(data);
    ```

     如果你没有截距或者想把它排除在计算之外,你需要使用一个不同的构造函数来创建 SimpleRegression 对象,如下所示:
    ```
    SimpleRegression regression = new SimpleRegression(false);
    ```

3. 找出截距、斜率,以及截距与斜率的标准误差。最后,关闭方法。

    ```
    System.out.println(regression.getIntercept());
    System.out.println(regression.getSlope());
      System.out.println(regression.getSlopeStdErr());
    }
    ```

完整代码整理如下:

```
import org.apache.commons.math3.stat.regression.SimpleRegression;

public class RegressionTest {
  public static void main(String[] args){
    double[][] data = { { 1, 3 }, {2, 5 }, {3, 7 }, {4, 14 }, {5, 11 }};
    RegressionTest test = new RegressionTest();
    test.calculateRegression(data);
  }
  public void calculateRegression(double[][] data){
    SimpleRegression regression = new SimpleRegression();
    regression.addData(data);
    System.out.println(regression.getIntercept());
    System.out.println(regression.getSlope());
```

```
        System.out.println(regression.getSlopeStdErr());
    }
}
```

 如果模型中观测值的数量少于两个，或者所有 x 值都一样，那么所有统计量都会返回 NaN。在获取统计量之后，如果你添加更多数据，就可以使用这些 getter 方法来得到更新后的统计量，而不必使用新实例。

3.9 计算普通最小二乘回归

OLSMultipleLinearRegression 支持普通最小二乘回归（ordinary least squares regression）来拟合线性模型 Y=X×b+u，其中 Y 是一个 n-vector 回归，X 是一个[n,k]矩阵，k 列称为回归子，b 是回归参数的 k-vector，u 是误差项或残差的 n-vector。

操作步骤

1. 创建一个名为 calculateOlsRegression 的方法，带有两个参数，一个是二维 double 型数组，另一个是一维 double 型数组。

    ```
    public void calculateOlsRegression(double[][] x, double[] y){
    ```

2. 创建一个 OLS 回归对象，并添加数据点 x 与 y。

    ```
    OLSMultipleLinearRegression regression = new
        OLSMultipleLinearRegression();
    regression.newSampleData(y, x);
    ```

3. 调用 OLSMultipleLinearRegression 类中的如下方法计算各种回归参数。如何使用这些信息取决于你面临的任务。最后，关闭方法。

    ```
    double[] beta = regression.estimateRegressionParameters();
    double[] residuals = regression.estimateResiduals();
    double[][] parametersVariance =
        regression.estimateRegressionParametersVariance();
    double regressandVariance =
        regression.estimateRegressandVariance();
    double rSquared = regression.calculateRSquared();
    double sigma = regression.estimateRegressionStandardError();
    }
    ```

4. 像下面这样,创建数据点 x 与 y。本示例中,我们使用的是固定数据,因此数组索引的初始值不是自动创建的。实际中,我们需要创建一个循环来创建 x 数组。

```
double[] y = new double[]{11.0, 12.0, 13.0, 14.0, 15.0, 16.0};
double[][] x = new double[6][];
x[0] = new double[]{0, 0, 0, 0, 0};
x[1] = new double[]{2.0, 0, 0, 0, 0};
x[2] = new double[]{0, 3.0, 0, 0, 0};
x[3] = new double[]{0, 0, 4.0, 0, 0};
x[4] = new double[]{0, 0, 0, 5.0, 0};
x[5] = new double[]{0, 0, 0, 0, 6.0};
```

示例的完整代码整理如下:

```
import org.apache.commons.math3.stat.regression.
  OLSMultipleLinearRegression;
public class OLSRegressionTest {
  public static void main(String[] args){
    double[] y = new double[]{11.0, 12.0, 13.0, 14.0, 15.0, 16.0};
    double[][] x = new double[6][];
    x[0] = new double[]{0, 0, 0, 0, 0};
    x[1] = new double[]{2.0, 0, 0, 0, 0};
    x[2] = new double[]{0, 3.0, 0, 0, 0};
    x[3] = new double[]{0, 0, 4.0, 0, 0};
    x[4] = new double[]{0, 0, 0, 5.0, 0};
    x[5] = new double[]{0, 0, 0, 0, 6.0};
    OLSRegressionTest test = new OLSRegressionTest();
    test.calculateOlsRegression(x, y);
  }
  public void calculateOlsRegression(double[][] x, double[] y){
    OLSMultipleLinearRegression regression = new
      OLSMultipleLinearRegression();
    regression.newSampleData(y, x);
    double[] beta = regression.estimateRegressionParameters();
    double[] residuals = regression.estimateResiduals();
    double[][] parametersVariance =
      regression.estimateRegressionParametersVariance();
    double regressandVariance =
      regression.estimateRegressandVariance();
    double rSquared = regression.calculateRSquared();
    double sigma = regression.estimateRegressionStandardError();
//在这里把这些值打印出来
```

    }
}

> 当出现如下两种情形时会抛出 IllegalArgumentException：一是输入数据的数组维数不匹配；二是数据数组未包含足够多的数据来评估模型。

3.10 计算广义最小二乘回归

本部分，我们将学习另外一种最小二乘回归，即广义最小二乘回归。GLSMultipleLinearRegression 实现了广义最小二乘法来拟合线性模型 Y=X×b+u。

操作步骤

1. 创建一个名为 calculateGlsRegression 的方法，它带有 3 个参数，第一个参数是二维 double 型数组，第二个参数是一维 double 数组，第三个参数是二维 double 数组，存放回归的 omega 参数。

    ```
    public void calculateGlsRegression(double[][] x, double[] y,
        double[][] omega){
    ```

2. 创建一个 GLS 回归对象、数据点与 omega 参数。

    ```
    GLSMultipleLinearRegression regression = new
        GLSMultipleLinearRegression();
    regression.newSampleData(y, x, omega);
    ```

3. 调用 GLSMultipleLinearRegression 类的方法，计算回归的各种统计量，最后关闭方法。

    ```
    double[] beta = regression.estimateRegressionParameters();
    double[] residuals = regression.estimateResiduals();
    double[][] parametersVariance =
        regression.estimateRegressionParametersVariance();
    double regressandVariance =
        regression.estimateRegressandVariance();
    double sigma = regression.estimateRegressionStandardError();
    }
    ```

4. 关于如何填充两个数组 x 与 y，请参考上一节内容。本示例中，除了数据点 x 与

y 之外，我们还需要 omega 值，可以把它们放入一个二维 double 型数组中，如下所示：

```
double[][] omega = new double[6][];
omega[0] = new double[]{1.1, 0, 0, 0, 0, 0};
omega[1] = new double[]{0, 2.2, 0, 0, 0, 0};
omega[2] = new double[]{0, 0, 3.3, 0, 0, 0};
omega[3] = new double[]{0, 0, 0, 4.4, 0, 0};
omega[4] = new double[]{0, 0, 0, 0, 5.5, 0};
omega[5] = new double[]{0, 0, 0, 0, 0, 6.6};
```

示例的完整代码如下，包含 calculateGlsRegression 方法、相应类及驱动方法。

```
import org.apache.commons.math3.stat.regression.
    GLSMultipleLinearRegression;
public class GLSRegressionTest {
    public static void main(String[] args){
        double[] y = new double[]{11.0, 12.0, 13.0, 14.0, 15.0, 16.0};
        double[][] x = new double[6][];
        x[0] = new double[]{0, 0, 0, 0, 0};
        x[1] = new double[]{2.0, 0, 0, 0, 0};
        x[2] = new double[]{0, 3.0, 0, 0, 0};
        x[3] = new double[]{0, 0, 4.0, 0, 0};
        x[4] = new double[]{0, 0, 0, 5.0, 0};
        x[5] = new double[]{0, 0, 0, 0, 6.0};
        double[][] omega = new double[6][];
        omega[0] = new double[]{1.1, 0, 0, 0, 0, 0};
        omega[1] = new double[]{0, 2.2, 0, 0, 0, 0};
        omega[2] = new double[]{0, 0, 3.3, 0, 0, 0};
        omega[3] = new double[]{0, 0, 0, 4.4, 0, 0};
        omega[4] = new double[]{0, 0, 0, 0, 5.5, 0};
        omega[5] = new double[]{0, 0, 0, 0, 0, 6.6};
        GLSRegressionTest test = new GLSRegressionTest();
        test.calculateGlsRegression(x, y, omega);
    }
    public void calculateGlsRegression(double[][] x, double[] y,
        double[][] omega){
        GLSMultipleLinearRegression regression = new
            GLSMultipleLinearRegression();
        regression.newSampleData(y, x, omega);
        double[] beta = regression.estimateRegressionParameters();
        double[] residuals = regression.estimateResiduals();
        double[][] parametersVariance =
```

```
            regression.estimateRegressionParametersVariance();
        double regressandVariance = 
            regression.estimateRegressandVariance();
        double sigma = regression.estimateRegressionStandardError();
//在这里打印数值
    }
}
```

3.11 计算两组数据点的协方差

无偏协方差的公式为 cov(X, Y) = sum [(xi - E(X))(yi -E(Y))] / (n - 1)，其中 E(X) 是指 X 的平均值，E(Y) 是指 Y 值的平均值。无偏估计（Non-bias-corrected estimates）使用 n 来代替 n-1。为了指定协方差是否为有偏估计（bias corrected），我们需要额外设置一个可选参数，即 biasCorrected，其默认设置为 true。

操作步骤

1. 创建一个名为 calculateCov 的方法，它带有两个一维数组，每个数组代表一组数据点。

   ```
   public void calculateCov(double[] x, double[] y){
   ```

2. 计算两组数据点的协方差，如下所示：

   ```
   double covariance = new Covariance().covariance(x, y, false);
   ```

   本例中，我们使用了无偏修正协方差，因此在 covariace() 方法中使用了 3 个参数。为了在两个 double 数组之间使用无偏协方差，移除第三个参数，如下：

   ```
   double covariance = new Covariance().covariance(x, y);
   ```

3. 根据你的需求，使用协方差，然后关闭方法。

   ```
   System.out.println(covariance);
   }
   ```

示例完整代码如下所示：

```
import org.apache.commons.math3.stat.correlation.Covariance;
```

```
public class CovarianceTest {
    public static void main(String[] args){
        double[] x = {43, 21, 25, 42, 57, 59};
        double[] y = {99, 65, 79, 75, 87, 81};
        CovarianceTest test = new CovarianceTest();
        test.calculateCov(x, y);
    }
    public void calculateCov(double[] x, double[] y){
        double covariance = new Covariance().covariance(x, y, false);//If
           false is removed, we get unbiased covariance
        System.out.println(covariance);
    }
}
```

3.12 为两组数据点计算皮尔逊相关系数

PearsonsCorrelation 用来计算相关系数，计算公式为 cor(X, Y) =sum[(xi - E(X))(yi - E(Y))] / [(n - 1)s(X)s(Y)]，其中，E(X) 与 E(Y) 分别是指 X 与 Y 的平均值，s(X) 与 s(Y) 是指标准差。

操作步骤

1. 创建一个名为 calculatePearson 的方法，其参数为两个 double 型数组，表示两组数据点。

   ```
   public void calculatePearson(double[] x, double[] y){
   ```

2. 创建一个 PearsonsCorrelation 对象。

   ```
   PearsonsCorrelation pCorrelation = new PearsonsCorrelation();
   ```

3. 计算两组数据点的相关系数。

   ```
   double cor = pCorrelation.correlation(x, y);
   ```

4. 根据你的需求，使用相关系数，最后关闭方法。

   ```
   System.out.println(cor);
   }
   ```

示例的完整代码如下所示：

```java
import org.apache.commons.math3.stat.correlation.PearsonsCorrelation;

public class PearsonTest {
   public static void main(String[] args){
      double[] x = {43, 21, 25, 42, 57, 59};
      double[] y = {99, 65, 79, 75, 87, 81};
      PearsonTest test = new PearsonTest();
      test.calculatePearson(x, y);
   }
   public void calculatePearson(double[] x, double[] y){
      PearsonsCorrelation pCorrelation = new PearsonsCorrelation();
      double cor = pCorrelation.correlation(x, y);
      System.out.println(cor);
   }
}
```

3.13　执行配对 t 检验

在 Apache Commons Math 所提供的众多标准统计显著性检验库中，我们将只使用其中几个来演示配对 t 检验、卡方检验、单因素方差分析、K-S 检验。各位可以执行其他显著性检验，相应代码将使用 `TestUtils` 类中的静态方法来执行检验。

Apache Commons Math 同时支持单样本与双样本 t 检验。而且，双样本检验既可以是配对的，也可以是非配对的。在执行非配对双样本检验时，亚群方差相等这个假设有无均可。

操作步骤

1. 创建一个名为 getTtest 的方法，其参数为两个 double 型数组，用来接收两组 double 值。我们将执行配对 t 检验，在这两组值之间找出任何统计显著性。

   ```java
   public void getTtest(double[] sample1, double[] sample2){
   ```

2. 调用 `pairedT()` 方法可以得到两种分布的 t 统计量。

   ```java
   System.out.println(TestUtils.pairedT(sample1, sample2));
   ```

3. 调用pairedTTest()方法可以得到配对t检验的p值。

    System.out.println(TestUtils.pairedTTest(sample1, sample2));

4. 最后，对于任意给定的置信区间或alpha值，两种分布的差异显著性可以像下面这样得到。

    System.out.println(TestUtils.pairedTTest(sample1, sample2, 0.05));

上面代码中，第三个参数设为0.05，表示我们想了解当把alpha水平设为0.05或者95%的置信区间时差异是否显著。

5. 最后，关闭方法。

}

示例完整代码如下所示：

```
import org.apache.commons.math3.stat.inference.TestUtils;
public class TTest {
  public static void main(String[] args){
      double[] sample1 = {43, 21, 25, 42, 57, 59};
      double[] sample2 = {99, 65, 79, 75, 87, 81};
      TTest test = new TTest();
      test.getTtest(sample1, sample2);
  }
  public void getTtest(double[] sample1, double[] sample2){
      System.out.println(TestUtils.pairedT(sample1, sample2));//t
        statistics
      System.out.println(TestUtils.pairedTTest(sample1, sample2));//p
        value
      System.out.println(TestUtils.pairedTTest(sample1, sample2,
        0.05));
  }
}
```

3.14 执行卡方检验

在对两组数据分布进行卡方检验时，其中一种分布叫观测分布（observed distribution），另一种分布叫预期分布（expected distribution）。

操作步骤

1. 创建一个名为 getChiSquare 的方法，其参数为两个分布。请注意，观测分布是一个 long 型数组，而预期分布是一个 double 数组。

   ```
   public void getChiSquare(long[] observed, double[] expected){
   ```

2. 获取卡方检验的 t 统计量，如下所示：

   ```
   System.out.println(TestUtils.chiSquare(expected, observed));
   ```

3. 采用类似方式，获取检验的 p 值，但所用的方法不同。

   ```
   System.out.println(TestUtils.chiSquareTest(expected, observed));
   ```

4. 对于一个给定的置信区间，我们还可以观察预期数据分布与观测数据分布之间的差异是否显著，如下所示：

   ```
   System.out.println(TestUtils.chiSquareTest(expected, observed, 0.05));
   ```

   上面代码中，我们把置信区间设为 95%，所以，把 chiSquareTest() 方法的第三个参数设置为 0.05。

5. 最后，关闭方法。

   ```
   }
   ```

   示例完整代码整理如下：

```
import org.apache.commons.math3.stat.inference.TestUtils;
public class ChiSquareTest {
  public static void main(String[] args){
     long[] observed = {43, 21, 25, 42, 57, 59};
     double[] expected = {99, 65, 79, 75, 87, 81};
     ChiSquareTest test = new ChiSquareTest();
     test.getChiSquare(observed, expected);
  }
```

```
public void getChiSquare(long[] observed, double[] expected){
    System.out.println(TestUtils.chiSquare(expected, observed));//t
        statistics
    System.out.println(TestUtils.chiSquareTest(expected,
        observed));//p value
    System.out.println(TestUtils.chiSquareTest(expected, observed,
        0.05));
    }
}
```

3.15 执行单因素方差分析（one-way ANOVA test）

ANOVA 是 Analysis of Variance（方差分析）的缩写。本部分，我们将学习如何使用 Java 来进行单因素方差分析，以判断 3 组（或 3 组以上）独立无关的数据点之间的差异是否显著。

操作步骤

1. 创建一个名为 calculateAnova 的方法，它接收各种数据分布。本示例中，我们将应用 ANOVA 来了解卡路里、肥胖、碳水化合物之间的关系。

   ```
   public void calculateAnova(double[] calorie, double[] fat,
       double[] carb, double[] control){
   ```

2. 创建一个 ArrayList，用来保存所有数据。作为方法参数接收的数据分布可以视作类别。本示例中，我们把它们命名为 classes。

   ```
   List<double[]> classes = new ArrayList<double[]>();
   ```

3. 依照顺序，把来自于四个类别的数据添加到 ArrayList。

   ```
   classes.add(calorie);
   classes.add(fat);
   classes.add(carb);
   classes.add(control);
   ```

4. 单因素方差分析的 F 值可以使用如下语句得到。

   ```
   System.out.println(TestUtils.oneWayAnovaFValue(classes));
   ```

5. 单因素方差分析的 p 值可以使用如下语句得到。

```
System.out.println(TestUtils.oneWayAnovaPValue(classes));
```

6. 最后，使用如下代码判断给定的四个类别的数据点之间的差异是否显著。

   ```
   System.out.println(TestUtils.oneWayAnovaTest(classes, 0.05));
   ```

7. 最后，使用右花括号关闭方法。

   ```
   }
   ```

单因素方差分析的完整代码如下：

```
import java.util.ArrayList;
import java.util.List;
import org.apache.commons.math3.stat.inference.TestUtils;
public class AnovaTest {
    public static void main(String[] args){
        double[] calorie = {8, 9, 6, 7, 3};
        double[] fat = {2, 4, 3, 5, 1};
        double[] carb = {3, 5, 4, 2, 3};
        double[] control = {2, 2, -1, 0, 3};
        AnovaTest test = new AnovaTest();
        test.calculateAnova(calorie, fat, carb, control);
    }
    public void calculateAnova(double[] calorie, double[] fat,
        double[]
    carb, double[] control){
        List<double[]> classes = new ArrayList<double[]>();
        classes.add(calorie);
        classes.add(fat);
        classes.add(carb);
        classes.add(control);
    System.out.println(TestUtils.oneWayAnovaFValue(classes));
    System.out.println(TestUtils.oneWayAnovaPValue(classes));
    System.out.println(TestUtils.oneWayAnovaTest(classes, 0.05));
    }
}
```

> 由 t 检验、卡方检验、ANOVA 检验返回的 p 值是准确的，它们基于 distribution 包中 t 分布、卡方分布、F 分布的数值近似。

3.16 执行 K-S 检验

本质上，K-S 检验（Kolmogorov-Smirnov test）是对连续的一维概率分布进行相等性检验（test of equality）。它是用来判断两组数据点间的差异是否明显的最流行的方法之一。

操作步骤

1. 创建一个名为 calculateKs 的方法，它接收两种不同的数据分布。我们将通过 K-S 检验来判断这两组数据分布之间的差异是否显著。

   ```
   public void calculateKs(double[] x, double[] y){
   ```

2. 检验中一个最重要的统计量是 d 统计量。它是一个 double 值，我们需要用它来计算检验的 p 值。

   ```
   double d = TestUtils.kolmogorovSmirnovStatistic(x, y);
   ```

3. 然后使用如下代码评估零假设，其值取自于单位正态分布。

   ```
   System.out.println(TestUtils.kolmogorovSmirnovTest(x, y,
       false));
   ```

4. 最后，使用如下代码得到显著性检验的 p 值。

   ```
   System.out.println(TestUtils.exactP(d, x.length, y.length,
       false));
   ```

示例的完整代码如下：

```
import org.apache.commons.math3.stat.inference.TestUtils;

public class KSTest {
  public static void main(String[] args){
    double[] x = {43, 21, 25, 42, 57, 59};
    double[] y = {99, 65, 79, 75, 87, 81};
    KSTest test = new KSTest();
    test.calculateKs(x, y);
  }
  public void calculateKs(double[] x, double[] y){
```

```
    double d = TestUtils.kolmogorovSmirnovStatistic(x, y);
System.out.println(TestUtils.kolmogorovSmirnovTest(x, y, false));
System.out.println(TestUtils.exactP(d, x.length, y.length,
    false));
}
}
```

到这里，本章内容就讲完了。借助 Apache Commons Math 库，你可以进行许多不同的统计分析。关于这个库的更多用法，请参考相应版本的说明文档：http://commons.apache.org/proper/commons-math/javadocs/api-3.6.1/index.html。

第 4 章 数据学习 I

本章涵盖内容如下：

- 创建与保存 ARFF 文件；
- 机器学习模型的交叉验证；
- 对未知测试数据分类；
- 使用过滤分类器对未知测试数据分类；
- 生成线性回归模型；
- 生成逻辑回归模型；
- 使用 K 均值算法对数据进行聚类；
- 依据类别对数据进行聚类处理；
- 从数据学习关联规则；
- 使用低层方法、过滤方法、元分类器方法选择特征/属性。

4.1 简介

在本章以及接下来的章节中，我们将讲解运用机器学习技术从数据学习模式的内容。这些模式至少是如下 3 种机器学习任务关注的焦点：分类、回归、聚类。分类任务是指从名义类预测一个值。与分类不同，回归模型试图从一个数字类预测一个值。而聚类则是一种基于邻近关系的数据点分组技术。

有很多基于 Java 的工具、工作平台、库、API 可以用来在前面提及的机器学习领域中进行研究与开发。**Weka**（**Waikato Environment of Knowledge Analysis**，怀卡托智能分析环境）是一种最流行的工具，它是一种免费的软件，其发行遵循 GNU 通用公共许可证。Weka 采用 Java 编写，拥有一系列非常棒的工具，可以用来做数据准备与过滤，并且包含经典的机器学习算法（参数设置可定制），以及强大的数字可视化工具。此外，它不仅提供了大量易用的 Java 库，还为非 Java 用户准备了简便易用的图形用户接口（GUI）。

本章讲解的重点是向各位演示如何使用 Weka 做常规的数据科学活动，比如为一个工具准备数据集，为不同类型的机器学习任务创建模型，以及评估模型性能。

请注意，本章各节的代码中不包含异常处理代码，因此在 catch 语句块中有意留空。异常处理完全取决于用户，即你的需求。

4.2 创建与保存 ARFF 文件

Weka 的原生文件格式叫"属性关联文件格式"（ARFF）。一个 ARFF 文件包含两个逻辑部分，第一部分是头部（header），包括 3 个必备的物理段，分别为关系名、属性或特征、数据类型与范围；第二部分是数据（data），其中用来生成机器学习模型的物理段也是必需的。ARFF 文件的头部格式如下：

```
% 1. Title: Iris Plants Database
%
% 2. Sources:
%      (a) Creator: R.A. Fisher
%      (b) Donor: Michael Marshall (MARSHALL%PLU@io.arc.nasa.gov)
%      (c) Date: July, 1988
%
@RELATION iris

@ATTRIBUTE sepallength      NUMERIC
@ATTRIBUTE sepalwidth       NUMERIC
@ATTRIBUTE petallength      NUMERIC
@ATTRIBUTE petalwidth       NUMERIC
@ATTRIBUTE class       {Iris-setosa,Iris-versicolor,Iris-virginica}
```

上面代码中，以 % 符号开始的各行是注释行，以 @RELATION 开始的行是关系名称行，而以@ATTRIBUTE 开始的各行表示特征或属性。在上面的示例中，关系的名称为 iris，数据集包含 5 个属性，其中前 4 个属性为 numeric 类型，最后一个属性为数据点所属的类别，它是一个 norminal 属性，带有 3 个类值。

ARFF 文件的数据段格式如下：

```
@DATA
5.1,3.5,1.4,0.2,Iris-setosa
4.9,3.0,1.4,0.2,Iris-setosa
4.7,3.2,1.3,0.2,Iris-setosa
4.6,3.1,1.5,0.2,Iris-setosa
5.0,3.6,1.4,0.2,Iris-setosa
5.4,3.9,1.7,0.4,Iris-setosa
4.6,3.4,1.4,0.3,Iris-setosa
```

上面示例中，数据段以@DATA 关键字开始，接着是以逗号分隔的属性值，这些属性值的顺序应该与属性段中的属性顺序保持一致。

请注意，关于@RELATION、@ATTRIBUTE、@DATA 的声明不区分大小写。

关于 ARFF 文件格式、Weka 所支持的属性类型、稀疏 ARFF 文件，若想了解更多的内容，请前往官网了解。

在正式开始之前，需要先做如下准备。

编写代码时，我们将使用 Eclipse IDE 开发环境，并且为了保证本章所有代码能够正常运行，还需要把 Weka JAR 文件添加到项目中。为此，做如下准备工作。

1. 下载 Weka，在下载页面中，你会看到针对 Windows、Mac OS X，以及其他操作系统平台（比如 Linux）的下载链接。请认真阅读下载说明，选择相应版本的 Weka 进行下载。

写作本书之时，面向开发者的 Weka 最新版本为 3.9.0。由于我使用的是 64 位的 Windows 操作系统，并且已经安装好了 JVM 1.8，所以在 Window 版块中选择不带 Java VM 的自解压可执行文件进行下载，如图 4-1 所示。

第 4 章 数据学习 I

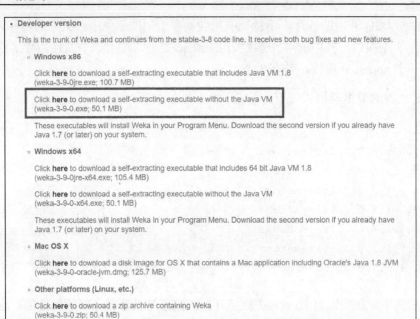

图 4-1

2. 下载完成后，双击可执行文件，根据屏幕提示进行安装即可。请注意，要选择 Weka 的完整版本进行安装。

3. 如图 4-2 所示，安装完成后，在运行软件之前，先去安装目录，找到 Weka 对应的 Java 文档文件（weka.jar），将其作为外部库添加到你的 Eclipse 项目中。

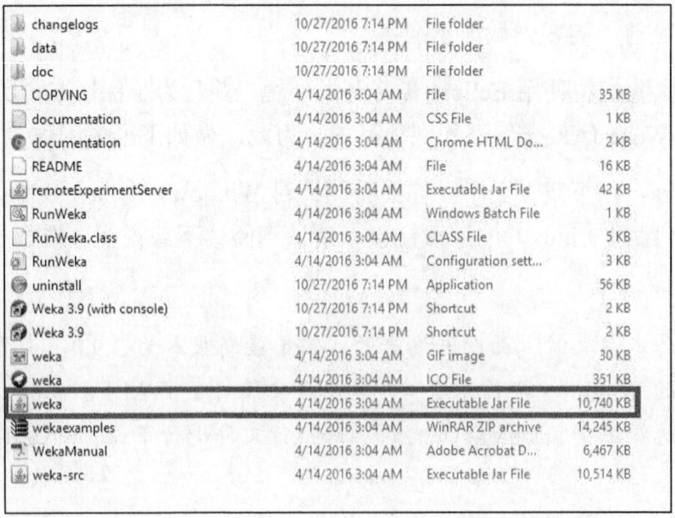

图 4-2

出于某种原因，你可能想下载旧版本的 Weka，你可以在官网中找到它们。请注意，旧版本中的许多方法可能已经过时了，因而不再提供支持。

操作步骤

1. 这里我们不会创建新方法，而是把所有代码放入 `main()` 方法中。首先创建 WekaArffTest 类及其 `main` 方法。

   ```
   public class WekaArffTest {
     public static void main(String[] args) throws Exception {
   ```

   请注意，由于 `main` 方法中所包含的代码与 Weka 库有关，因此有可能会抛出异常。

2. 创建两个 `ArrayList`，第一个用来保存属性，第二个用来保存类值。因此，第一个 `ArrayList` 的泛型将是 Attribute（其实它是 Weka 中一个模型属性类），而第二个 `ArrayList` 的泛型是 String，表示类标签。

   ```
   ArrayList<Attribute>     attributes;
   ArrayList<String>        classVals;
   ```

3. 接着，创建一个 Instances 对象。这个对象将用来为 ARFF 文件的 `@DATA` 部分中的实例建模。`@DATA` 部分的每一行就是一个实例。

   ```
   Instances     data;
   ```

4. 创建一个 double 型数组，该数组用来保存属性值。

   ```
   double[]      values;
   ```

5. 接下来，该创建属性了。我们将在 ARFF 文件中创建 `@ATTRIBUTE` 部分。首先，对属性进行实例化。

   ```
   attributes = new ArrayList<Attribute>();
   ```

6. 接着，创建一个名为 age 的数值型属性，并且将其添加到属性 `ArrayList` 之中。

   ```
   attributes.add(new Attribute("age"));
   ```

7. 然后，创建一个名为 name 的字符串属性，并且将其添加到属性 `ArrayList` 之中。

但在此之前,我们先要创建一个空的 String 类型的 `ArrayList`,并把 NULL 值赋给它。这个空的 `ArrayList` 将用在 Attribute 类的构造函数中,以表示 name 是一个 String 类型的属性,而不像名义型的类属性。

```
ArrayList<String> empty = null;
attributes.add(new Attribute("name", empty));
```

8. Weka 还支持日期类型的属性。接下来,我们将创建一个 dob 属性,用来表示出生日期。

```
attributes.add(new Attribute("dob", "yyyy-MM-dd"));
```

9. 然后,我们对类值 `ArrayList` 进行实例化,并且创建 5 个类值:class1、class2、class3、class4、class5。

```
classVals = new ArrayList<String>();
for (int i = 0; i < 5; i++){
    classVals.add("class" + (i + 1));
}
```

10. 接下来,使用这些类值创建一个属性,并将其添加到我们的属性 `ArrayList` 中。

```
Attribute classVal = new Attribute("class", classVals);
attributes.add(classVal);
```

11. 借助这行代码,我们就为 ARFF 文件创建好了 @ATTRIBUTE 部分。接着,我们要填充 ARFF 文件的 @DATA 部分。

12. 首先,创建一个 Instances 对象,接收关系名 `MyRelation`(该参数对应于 ARFF 文件的 @RELATION 部分)与所有属性。

```
data = new Instances("MyRelation", attributes, 0);
```

13. 使用之前创建的 double 型数组为 4 个属性生成 4 个值。把年龄、名字、出生日期、类值(本示例中它不重要,随机选择即可)赋给数组各个元素。

```
values = new double[data.numAttributes()];
values[0] = 35;
values[1] = data.attribute(1).addStringValue("John Doe");
values[2] = data.attribute(2).parseDate("1981-01-20");
values[3] = classVals.indexOf("class3");
```

14. 然后，我们把这些值添加到数据段。

    ```
    data.add(new DenseInstance(1.0, values));
    ```

15. 采用类似方法，我们为数据段再创建一个实例，如下所示：

    ```
    values = new double[data.numAttributes()];
    values[0] = 30;
    values[1] = data.attribute(1).addStringValue("Harry Potter");
    values[2] = data.attribute(2).parseDate("1986-07-05");
    values[3] = classVals.indexOf("class1");
    data.add(new DenseInstance(1.0, values));
    ```

16. 如果你想把 ARFF 文件保存到某个指定的地方，请添加如下代码段：

    ```
    BufferedWriter writer = new BufferedWriter(new
      FileWriter("c:/training.arff"));
    writer.write(data.toString());
    writer.close();
    ```

17. 把我们刚刚创建的 ARFF 文件的所有内容显示在控制台中，代码如下所示：

    ```
    System.out.println(data);
    ```

18. 最后，关闭方法与类。

    ```
        }
    }
    ```

示例的完整代码整理如下：

```
import java.io.BufferedWriter;
import java.io.FileWriter;
import java.util.ArrayList;

import weka.core.Attribute;
import weka.core.DenseInstance;
import weka.core.Instances;

public class WekaArffTest {
    public static void main(String[] args) throws Exception {
        ArrayList<Attribute>    attributes;
        ArrayList<String>       classVals;
```

```
Instances        data;
double[]         values;
//创建属性
attributes = new ArrayList<Attribute>();
//数值属性
attributes.add(new Attribute("age"));
//字符串属性
ArrayList<String> empty = null;
attributes.add(new Attribute("name", empty));
//日期属性
attributes.add(new Attribute("dob", "yyyy-MM-dd"));
classVals = new ArrayList<String>();
for (int i = 0; i < 5; i++){
   classVals.add("class" + (i + 1));
}
Attribute classVal = new Attribute("class", classVals);
attributes.add(classVal);

//创建实例对象
data = new Instances("MyRelation", attributes, 0);

//填充日期
//第一个实例
values = new double[data.numAttributes()];
values[0] = 35;
values[1] = data.attribute(1).addStringValue("John Doe");
values[2] = data.attribute(2).parseDate("1981-01-20");
values[3] = classVals.indexOf("class3");

//添加
data.add(new DenseInstance(1.0, values));

//第二个实例
values = new double[data.numAttributes()];
values[0] = 30;
values[1] = data.attribute(1).addStringValue("Harry Potter");
values[2] = data.attribute(2).parseDate("1986-07-05");
values[3] = classVals.indexOf("class1");

//添加
data.add(new DenseInstance(1.0, values));

//把 ARFF 文件写到磁盘
```

```
        BufferedWriter writer = new BufferedWriter(new
          FileWriter("c:/training.arff"));
        writer.write(data.toString());
        writer.close();
        //输出数据
        System.out.println(data);
    }
}
```

运行以上代码,输出如下结果:

```
@relation MyRelation

@attribute age numeric
@attribute name string
@attribute dob date yyyy-MM-dd
@attribute class {class1,class2,class3,class4,class5}

@data
35,'John Doe',1981-01-20,class3
30,'Harry Potter',1986-07-05,class1
```

4.3　对机器学习模型进行交叉验证

本部分,我们将创建 4 个方法,用来做 4 件不同的事情,第一个方法用来加载 ARFF 文件(假设 ARFF 文件已经创建好,并且保存于某个地方),第二个方法用来读取 ARFF 文件中的数据,并生成一个机器学习模型(我们任意选择了朴素贝叶斯模型),第三个方法使用序列化来存储模型,最后一个方法使用 10 折交叉验证评估基于 ARFF 文件的模型。

操作步骤

1. 创建两个 Instances 变量,第一个用来保存 iris 数据集的所有实例(iris ARFF 数据集存在于 Weka 安装目录的 data 文件夹中),第二个变量是 NaiveBayes 分类器。

   ```
   Instances iris = null;
   NaiveBayes nb;
   ```

2. 第一个方法将使用 DataSource 类来加载 iris ARFF 文件,调用这个类的 getDataSet() 方法来读取内容,并且设置类属性的位置。如果使用 notepad 打开

iris.arff 数据集，你将会看到类属性是最后一个属性，这只是一个习惯，不是强制规则。iris.setClassIndex(iris.numAttributes() - 1)用来把最后一个属性设置为类属性。在 Weka 的任意一个分类任务中，这是非常重要的。

```
public void loadArff(String arffInput){
  DataSource source = null;
  try {
    source = new DataSource(arffInput);
    iris = source.getDataSet();
    if (iris.classIndex() == -1)
      iris.setClassIndex(iris.numAttributes() - 1);
  } catch (Exception e1) {
  }
}
```

3. 第二个方法将使用 NaiveBayes 类的 buildClassifier(dataset) 方法创建一个基于 iris 数据集的朴素贝叶斯分类器。

```
public void generateModel(){
  nb = new NaiveBayes();
  try {
    nb.buildClassifier(iris);
  } catch (Exception e) {
  }
}
```

4. Weka 提供了一个工具类，用来把使用 Weka 生成的模型保存起来。以后我们可以使用这些保存下来的模型对那些新的、未打过标签的测试数据进行分类。这里我们要使用的工具类就是 Weka 的 SerializationHelper 类，它有一个名称为 write 的特定方法，它有两个参数，第一个是模型的保存路径，第二个是要保存的模型。

```
public void saveModel(String modelPath){
  try {
    weka.core.SerializationHelper.write(modelPath, nb);
  } catch (Exception e) {
  }
}
```

5. 最后一个方法使用 iris 数据集对模型进行交叉验证，以评估模型的性能。为此，我们将使用 10 折交叉验证。这种模型性能评估技术非常流行，并且当我们拥有少量数据时，它非常有用。但是，它本身也带有一些限制。关于这个方法优缺点的讨论已超出本书范畴，在此不谈，感兴趣的朋友，请前往 https://en.wikipedia.org/wiki/Cross-validation_(statistics) 了解更多的细节。

```java
public void crossValidate(){
  Evaluation eval = null;
  try {
    eval = new Evaluation(iris);
    eval.crossValidateModel(nb, iris, 10, new Random(1));
    System.out.println(eval.toSummaryString());
  } catch (Exception e1) {
  }
}
```

在 eval.crossValidateModel(nb, iris, 10, new Random(1)) 代码行中，前两个参数分别是所使用的模型与数据集，第三个参数表示使用的是 10 折交叉验证，最后一个参数把随机化引入处理过程中，这是非常重要的一个参数，因为大部分情况下数据集中的数据实例都不是随机的。

示例的完整可执行代码整理如下：

```java
import java.util.Random;

import weka.classifiers.Evaluation;
import weka.classifiers.bayes.NaiveBayes;
import weka.core.Instances;
import weka.core.converters.ConverterUtils.DataSource;

public class WekaCVTest {
  Instances iris = null;
  NaiveBayes nb;

  public void loadArff(String arffInput){
    DataSource source = null;
    try {
      source = new DataSource(arffInput);
      iris = source.getDataSet();
      if (iris.classIndex() == -1)
        iris.setClassIndex(iris.numAttributes() - 1);
    } catch (Exception e1) {
    }
  }
```

```java
    public void generateModel(){
        nb = new NaiveBayes();
        try {
            nb.buildClassifier(iris);
        } catch (Exception e) {
        }
    }

    public void saveModel(String modelPath){
        try {
            weka.core.SerializationHelper.write(modelPath, nb);
        } catch (Exception e) {
        }
    }
    public void crossValidate(){
        Evaluation eval = null;
        try {
            eval = new Evaluation(iris);
            eval.crossValidateModel(nb, iris, 10, new Random(1));
            System.out.println(eval.toSummaryString());
        } catch (Exception e1) {
        }
    }

    public static void main(String[] args){
        WekaCVTest test = new WekaCVTest();
        test.loadArff("C:/Program Files/Weka-3-6/data/iris.arff");
        test.generateModel();
        test.saveModel("c:/nb.model");
        test.crossValidate();
    }
}
```

运行上面代码，得到如下输出结果：

```
Correctly Classified Instances          144             96      %
Incorrectly Classified Instances          6              4      %
Kappa statistic                         0.94
Mean absolute error                     0.0342
Root mean squared error                 0.155
Relative absolute error                 7.6997  %
```

```
Root relative squared error          32.8794 %
Total Number of Instances            150
```

本示例中，我们把机器学习模型保存起来了。如果需要加载一个模型，你需要知道这个模型是使用哪种学习算法（示例中使用的是朴素贝叶斯算法）创建的，这样你才能把模型加载到相应的学习算法对象中。你可以使用如下方法加载模型。

```
public void loadModel(String modelPath){
try {
nb = (NaiveBayes)
weka.core.SerializationHelper.read(modelPath);
} catch (Exception e) {
}
}
```

4.4 对新的测试数据进行分类

经典的有监督机器学习分类任务是基于带标签的训练实例来训练分类器，然后把分类器用在新的测试实例上。这里需要记住的重要一点是训练集中的属性数量、类型、名称，以及取值范围（如果它们是常规的名义属性或名义类属性）必须与测试数据集中的那些完全一样。

准备工作

在 Weka 中，训练数据集与测试数据集之间可能存在关键的不同。在测试部分，ARFF 文件的 @DATA 部分看上去与一个 ARFF 文件的 @DATA 部分类似。它可以包含如下属性值与类标签。

```
@DATA
    5.1,3.5,1.4,0.2,Iris-setosa
    4.9,3.0,1.4,0.2,Iris-setosa
    4.7,3.2,1.3,0.2,Iris-setosa
```

当把一个分类器应用到这些带标签的测试数据时，分类器在预测一个实例的类别时会忽略类标签。另外，请注意，如果测试数据带有标签，你可以把分类器预测出的标签与真实的标签进行比较。这让你有机会为分类器生成评估指标。然而，大部分情况下，你的测试数据不带有任何类别信息，在分类器对训练数据进行学习之后，就可以用它对这些测试数据进行预测，并打上类别标签。测试数据集的 `@DATA` 部分类似于下面这个样子，其中类标签是未知的，用问号（？）表示。

```
@DATA
5.1,3.5,1.4,0.2,?
4.9,3.0,1.4,0.2,?
4.7,3.2,1.3,0.2,?
```

Weka 的 Data 目录中不包含这样的测试文件。因此，需要你仿照 `iris.arff` 文件自己创建测试文件。首先复制如下内容，打开 notepad，把它们放入一个文本文件中，然后把这个文本文件保存到你的文件系统中，得到 `iris-test.arff` 文件（假设保存在 C:/r 目录之下）。

```
@RELATION iris-test

@ATTRIBUTE sepallength   REAL
@ATTRIBUTE sepalwidth    REAL
@ATTRIBUTE petallength   REAL
@ATTRIBUTE petalwidth    REAL
@ATTRIBUTE class {Iris-setosa,Iris-versicolor,Iris-virginica}
@DATA
3.1,1.2,1.2,0.5,?
2.3,2.3,2.3,0.3,?
4.2,4.4,2.1,0.2,?
3.1,2.5,1.0,0.2,?
2.8,1.6,2.0,0.2,?
3.0,2.6,3.3,0.3,?
4.5,2.0,3.4,0.1,?
5.3,2.0,3.1,0.2,?
3.2,1.3,2.1,0.3,?
2.1,6.4,1.2,0.1,?
```

操作步骤

1. 创建如下实例变量：

   ```
   NaiveBayes nb;
   Instances train, test, labeled;
   ```

为了增加一点挑战难度，获得一个好的学习机会，我们将加载前面创建并保存过的模型，并把它赋给我们的 `NaiveBayes` 分类器。这个分类器将应用到未打标签的测试实例上。把测试实例作为带标签的实例复制给分类器。不需要改变测试实例，分类器预测的分类标签将作为类别标签指派给各个带标签的实例。

2. 首先，创建一个方法，用来加载预先创建好并保存下来的模型。其实，我们可以加载在前面一节创建并保存的分类器。

```
public void loadModel(String modelPath){
  try {
    nb = (NaiveBayes)
      weka.core.SerializationHelper.read(modelPath);
  } catch (Exception e) {
  }
}
```

3. 然后，我们需要读取训练与测试数据集。作为训练数据集，我们将使用 Weka 的 Data 目录下的 `iris.arff` 文件。并且，我们将把前面创建好的 `iris-test.arff` 文件用作测试文件。

```
public void loadDatasets(String training, String testing){
```

4. 前面在读取训练数据集时，我们使用了 Weka 的 `DataSource` 类。使用这个类的最大好处是它可以处理 Weka 支持的所有类型的文件。当然，Weka 用户也可以使用 Java 的 `BufferedReader` 类来读取数据集的内容。本部分，为了向大家介绍这种读取数据集的方式，我们将使用 `BufferedReader` 类来代替 `DataSource` 类。

   我们将把 ARFF 文件看作普通的文件，使用一个 `BufferedReader` 类型的 reader 指向训练数据集。然后使用 Weka 的 `Instances` 类的构造函数，把 reader 作为参数传入以便创建训练实例。最后，我们把最后一个属性设置成这个数据集的类属性。

```
BufferedReader reader = null;
  try {
    reader = new BufferedReader(new FileReader(training));
    train = new Instances (reader);
    train.setClassIndex(train.numAttributes() -1);
  } catch (IOException e) {
}
```

5. 采用类似方式，读取测试数据集。

```
try {
    reader = new BufferedReader(new FileReader(testing));
    test = new Instances (reader);
    test.setClassIndex(train.numAttributes() -1);
} catch (IOException e) {
}
```

请注意，这里我们不必创建新的 `BufferedReader` 对象。

6. 最后，关闭打开的 `BufferedReader` 对象，结束方法。

```
try {
    reader.close();
} catch (IOException e) {
}
}
```

7. 在接下来的方法中，我们使用训练数据创建一个 `NaiveBayes` 分类器，并且把这个分类器应用到测试数据集中那些之前不曾见过且未打标签的实例上。这个方法还会显示 `NaiveBayes` 分类器所预测的类值的概率。

```
public void classify(){
    try {
        nb.buildClassifier(train);
    } catch (Exception e) {
    }
}
```

8. 我们将创建带标签的实例，它们是测试实例的副本。使用上一步所创建的分类器进行预测，所得到的标签会指派给这些实例，保持测试实例不变。

```
labeled = new Instances(test);
```

9. 接着，为测试集的每个实例，创建一个类标签，它是一个 double 变量。然后调用 Naive Bayes 的 `classifyInstance()` 方法，它接收一个实例作为参数。类标签将指派给类标签变量，并且这个变量的值会被指派为带标签的实例中这个特定实例的类标签。换言之，在带标签的实例中，测试实例的？值将被 Naive Bayes 预测的值所取代。

```
for (int i = 0; i < test.numInstances(); i++) {
    double clsLabel;
```

```
        try {
            clsLabel = nb.classifyInstance(test.instance(i));
            labeled.instance(i).setClassValue(clsLabel);
            double[] predictionOutput =
               nb.distributionForInstance(test.instance(i));
            double predictionProbability = predictionOutput[1];
            System.out.println(predictionProbability);
        } catch (Exception e) {
        }
    }
```

10. 最后，我们把打过标签的测试数据集（即 `labeled`）写入文件系统中。

```
public void writeArff(String outArff){
    BufferedWriter writer;
    try {
        writer = new BufferedWriter(new FileWriter(outArff));
        writer.write(labeled.toString());
        writer.close();
    } catch (IOException e) {
    }
}
```

示例的完整可执行代码整理如下：

```
import java.io.BufferedReader;
import java.io.BufferedWriter;
import java.io.FileReader;
import java.io.FileWriter;
import java.io.IOException;

import weka.classifiers.bayes.NaiveBayes;
import weka.core.Instances;

public class WekaTrainTest {
   NaiveBayes nb;
   Instances train, test, labeled;
   public void loadModel(String modelPath){
      try {
         nb = (NaiveBayes)
            weka.core.SerializationHelper.read(modelPath);
      } catch (Exception e) {
      }
```

```java
    }
    public void loadDatasets(String training, String testing){
        BufferedReader reader = null;
        try {
            reader = new BufferedReader(new FileReader(training));
            train = new Instances (reader);
            train.setClassIndex(train.numAttributes() -1);
        } catch (IOException e) {
        }

        try {
            reader = new BufferedReader(new FileReader(testing));
            test = new Instances (reader);
            test.setClassIndex(train.numAttributes() -1);
        } catch (IOException e) {
        }

        try {
            reader.close();
        } catch (IOException e) {
        }
    }

    public void classify(){
        try {
            nb.buildClassifier(train);
        } catch (Exception e) {
        }

        labeled = new Instances(test);

        for (int i = 0; i < test.numInstances(); i++) {
            double clsLabel;
            try {
                clsLabel = nb.classifyInstance(test.instance(i));
                labeled.instance(i).setClassValue(clsLabel);
                double[] predictionOutput =
                    nb.distributionForInstance(test.instance(i));
                double predictionProbability = predictionOutput[1];
                System.out.println(predictionProbability);
            } catch (Exception e) {
            }
        }
```

```
    }
    public void writeArff(String outArff){
        BufferedWriter writer;
        try {
            writer = new BufferedWriter(new FileWriter(outArff));
            writer.write(labeled.toString());
            writer.close();
        } catch (IOException e) {
        }
    }
    public static void main(String[] args) throws Exception{
        WekaTrainTest test = new WekaTrainTest();
        test.loadModel("path to your Naive Bayes Model");
        test.loadDatasets("path to iris.arff dataset", "path to iristest.
           arff dataset");
        test.classify();
        test.writeArff("path to your output ARFF file");
    }
}
```

执行上面的代码,在控制台中,你将会看到模型预测的概率值。

```
5.032582653870928E-13
2.1050052853672135E-4
5.177104804026096E-16
1.2459904922893976E-16
3.1771015903129274E-10
0.9999993509430146
0.999999944638627
0.9999999844862647
3.449759371835354E-8
4.0178483420981394E-77
```

打开代码所生成的 ARFF 文件,里面包含着先前未知实例的类值,你应该会看到如下内容。

```
@relation iris-test

@attribute sepallength numeric
@attribute sepalwidth numeric
@attribute petallength numeric
@attribute petalwidth numeric
```

```
@attribute class {Iris-setosa,Iris-versicolor,Iris-virginica}

@data
3.1,1.2,1.2,0.5,Iris-setosa
2.3,2.3,2.3,0.3,Iris-setosa
4.2,4.4,2.1,0.2,Iris-setosa
3.1,2.5,1,0.2,Iris-setosa
2.8,1.6,2,0.2,Iris-setosa
3,2.6,3.3,0.3,Iris-versicolor
4.5,2,3.4,0.1,Iris-versicolor
5.3,2,3.1,0.2,Iris-versicolor
3.2,1.3,2.1,0.3,Iris-setosa
2.1,6.4,1.2,0.1,Iris-setosa
```

4.5 使用过滤分类器对新测试数据分类

许多时候，在开发分类器之前，我们需要使用过滤器进行一系列的处理，比如移除、转换、离散化，以及添加属性，移除误分类实例，对实例进行随机化或标准化处理等。为此，通常的方法是使用 Weka 的 Filter 类，调用这个类的方法进行一系列的过滤处理。此外，Weka 中还有一个名为 FilteredClassifier 的类，这个类用来在经过过滤器过滤的数据上运行分类器。

本部分，我们将学习如何同时使用过滤器与分类器对先前未见过的测试样例进行分类。

操作步骤

1. 这次，我们将使用随机森林分类器，并且把安装在文件系统中的 Weka 文件夹中 Data 目录下的 weather.nominal.arff 用作数据集。

   下面两个是我们的实例变量。

   ```
   Instances weather = null;
   RandomForest rf;
   ```

2. 接下来，我们需要创建一个方法，用来加载数据集。并且通过驱动方法把 weather.nominal.arff 文件的路径传递给这个方法。借助 Weka 的 DataSource 类，读取 weather.nominal.arff 文件的数据，把数据集的最后一个属性设置为类属性。

```
public void loadArff(String arffInput){
  DataSource source = null;
  try {
    source = new DataSource(arffInput);
    weather = source.getDataSet();
    weather.setClassIndex(iris.numAttributes() - 1);
  } catch (Exception e1) {
  }
}
```

3. 接下来，我们创建本部分最关键的方法。

```
public void buildFilteredClassifier(){
```

4. 在创建这个方法之前，我们先要创建一个随机森林分类器。

```
rf = new RandomForest();
```

5. 然后创建一个过滤器，用来从 weather.nominal.arff 文件移除特定的属性。为此，我们会使用 Weka 的 Remove 类。下面代码创建的过滤器将用来移除数据集的第一个属性。

```
Remove rm = new Remove();
rm.setAttributeIndices("1");
```

6. 接下来，创建一个 FilteredClassifier，添加在上一步中创建的过滤器，添加 RandomForest 分类器。

```
FilteredClassifier fc = new FilteredClassifier();
fc.setFilter(rm);
fc.setClassifier(rf);
```

7. 借助过滤器和分类器，我们可以根据名义天气数据集创建一个随机森林分类器。然后，分类器为天气数据集的每个实例预测类值。在 try 语句块中，我们将把实例的实际值与预测值打印出来。

```
try{
  fc.buildClassifier(weather);
  for (int i = 0; i < iris.numInstances(); i++){
    double pred = fc.classifyInstance(weather.instance(i));
```

```java
            System.out.print("given value: " +
               weather.classAttribute().value((int)
               weather.instance(i).classValue()));
            System.out.println("---predicted value: " +
               weather.classAttribute().value((int) pred));
         }
      } catch (Exception e) {
      }
   }
```

示例的完整代码整理如下：

```java
import weka.classifiers.meta.FilteredClassifier;
import weka.classifiers.trees.RandomForest;
import weka.core.Instances;
import weka.core.converters.ConverterUtils.DataSource;
import weka.filters.unsupervised.attribute.Remove;

public class WekaFilteredClassifierTest {
   Instances weather = null;
   RandomForest rf;

   public void loadArff(String arffInput){
      DataSource source = null;
      try {
         source = new DataSource(arffInput);
         weather = source.getDataSet();
         weather.setClassIndex(weather.numAttributes() - 1);
      } catch (Exception e1) {
      }
   }

   public void buildFilteredClassifier(){
      rf = new RandomForest();
      Remove rm = new Remove();
      rm.setAttributeIndices("1");
      FilteredClassifier fc = new FilteredClassifier();
      fc.setFilter(rm);
      fc.setClassifier(rf);
      try{
         fc.buildClassifier(weather);
         for (int i = 0; i < weather.numInstances(); i++){
            double pred = fc.classifyInstance(weather.instance(i));
```

```
            System.out.print("given value: " +
              weather.classAttribute().value((int)
                weather.instance(i).classValue()));
            System.out.println("---predicted value: " +
              weather.classAttribute().value((int) pred));
        }
    } catch (Exception e) {
    }
}
public static void main(String[] args){
    WekaFilteredClassifierTest test = new
      WekaFilteredClassifierTest();
    test.loadArff("C:/Program Files/Weka-3-
      6/data/weather.nominal.arff");
    test.buildFilteredClassifier();
}
}
```

运行上面的代码，输出如下结果：

```
given value: no---predicted value: yes
given value: no---predicted value: no
given value: yes---predicted value: yes
given value: yes---predicted value: yes
given value: yes---predicted value: yes
given value: no---predicted value: yes
given value: yes---predicted value: yes
given value: no---predicted value: yes
given value: yes---predicted value: yes
given value: yes---predicted value: yes
given value: yes---predicted value: yes
given value: yes---predicted value: yes
given value: yes---predicted value: yes
given value: no---predicted value: yes
```

4.6　创建线性回归模型

大多数线性回归模型遵循一个通用模式，即会有许多独立变量共同产生一个非独立变量的结果。比如，我们可以创建一个回归模型，用来根据房屋的不同属性/特征［大部分是真实数值，比如面积（平方英尺）、卧室数量、洗手间数量、位置的重要性等］来预测房屋的价格。

本部分，我们将使用 Weka 的线性回归分类器来创建回归模型。

操作步骤

1. 这里，我们要创建的线性回归模型是基于 `cpu.arff` 数据集的，你可以在 Weka 的安装目录下的 `data` 目录中找到这个数据集。

   在代码中我们将创建两个实例变量，第一个变量用来保存 `cpu.arff` 文件的数据实例，第二个变量是线性回归分类器。

   ```
   Instances cpu = null;
   LinearRegression lReg ;
   ```

2. 接着，我们创建一个方法，用来加载 ARFF 文件，并把 ARFF 文件的最后一个属性指定为类别属性。

   ```
   public void loadArff(String arffInput){
     DataSource source = null;
     try {
       source = new DataSource(arffInput);
       cpu = source.getDataSet();
       cpu.setClassIndex(cpu.numAttributes() - 1);
     } catch (Exception e1) {
     }
   }
   ```

3. 然后，我们将创建一个方法，用来创建线性回归模型。为此，我们需要调用线性回归变量的 `buildClassifier()` 方法。我们可以把这个模型作为参数直接传递给 `System.out.println()`。

   ```
   public void buildRegression(){
     lReg = new LinearRegression();
     try {
       lReg.buildClassifier(cpu);
     } catch (Exception e) {
     }
     System.out.println(lReg);
   }
   ```

示例的完整代码整理如下：

```
import weka.classifiers.functions.LinearRegression;
import weka.core.Instances;
import weka.core.converters.ConverterUtils.DataSource;

public class WekaLinearRegressionTest {
   Instances cpu = null;
   LinearRegression lReg ;

   public void loadArff(String arffInput){
      DataSource source = null;
      try {
         source = new DataSource(arffInput);
         cpu = source.getDataSet();
         cpu.setClassIndex(cpu.numAttributes() - 1);
      } catch (Exception e1) {
      }
   }

   public void buildRegression(){
      lReg = new LinearRegression();
      try {
         lReg.buildClassifier(cpu);
      } catch (Exception e) {
      }
      System.out.println(lReg);
   }

   public static void main(String[] args) throws Exception{
      WekaLinearRegressionTest test = new WekaLinearRegressionTest();
      test.loadArff("path to the cpu.arff file");
      test.buildRegression();
   }
}
```

运行上面代码，输出如下结果：

```
Linear Regression Model

class =

      0.0491 * MYCT +
      0.0152 * MMIN +
```

```
      0.0056 * MMAX +
      0.6298 * CACH +
      1.4599 * CHMAX +
 -56.075
```

4.7 创建逻辑回归模型

Weka 中有一个名为 Logistic 的类,用来创建与使用带有岭估计的多项逻辑回归模型。尽管原始的逻辑回归无法处理实例权重,但是 Weka 中的算法进行了相应修改,因而可以处理实例权重。

本部分,我们将使用 Weka 基于 iris 数据集创建逻辑回归模型。

操作步骤

1. 首先,我们根据 iris 数据集创建逻辑回归模型,在 Weka 文件夹的 data 目录中,你可以找到这个数据集。

   在代码中,我们创建了两个实例变量,一个用来保存 iris 数据集的数据实例,另一是逻辑回归分类器。

   ```
   Instances iris = null;
   Logistic logReg ;
   ```

2. 接着,创建一个方法,用来加载与读取数据集,以及指派类属性(iris.arff 文件的最后一个属性)。

   ```
   public void loadArff(String arffInput){
     DataSource source = null;
     try {
       source = new DataSource(arffInput);
       iris = source.getDataSet();
       iris.setClassIndex(iris.numAttributes() - 1);
     } catch (Exception e1) {
     }
   }
   ```

3. 然后,创建本部分最重要的一个方法,即从 iris 数据集创建一个逻辑回归分类器。

   ```
   public void buildRegression(){
   ```

```
      logReg = new Logistic();
      try {
        logReg.buildClassifier(iris);
      } catch (Exception e) {
      }
      System.out.println(logReg);
   }
```

示例完整的可执行代码如下：

```
import weka.classifiers.functions.Logistic;
import weka.core.Instances;
import weka.core.converters.ConverterUtils.DataSource;

public class WekaLogisticRegressionTest {
   Instances iris = null;
   Logistic logReg ;

   public void loadArff(String arffInput){
      DataSource source = null;
      try {
         source = new DataSource(arffInput);
         iris = source.getDataSet();
         iris.setClassIndex(iris.numAttributes() - 1);
      } catch (Exception e1) {
      }
   }

   public void buildRegression(){
      logReg = new Logistic();

      try {
        logReg.buildClassifier(iris);
      } catch (Exception e) {
      }
      System.out.println(logReg);
   }

   public static void main(String[] args) throws Exception{
      WekaLogisticRegressionTest test = new
         WekaLogisticRegressionTest();
      test.loadArff("path to the iris.arff file ");
      test.buildRegression();
```

    }
}

运行上面代码，得到如下输出结果：

```
Logistic Regression with ridge parameter of 1.0E-8
Coefficients...
                          Class
Variable          Iris-setosa    Iris-versicolor
===================================================
sepallength             21.8065           2.4652
sepalwidth               4.5648           6.6809
petallength            -26.3083          -9.4293
petalwidth             -43.887          -18.2859
Intercept                8.1743          42.637


Odds Ratios...
                          Class
Variable          Iris-setosa    Iris-versicolor
===================================================
sepallength     2954196659.8892          11.7653
sepalwidth              96.0426         797.0304
petallength                   0           0.0001
petalwidth                    0                0
```

 对上面输出结果的解读已经超出本书的讨论范畴，感兴趣的朋友请参阅 Stack Overflow 官网中的相关讨论。

4.8 使用 K 均值算法对数据点进行聚类

本部分，我们将学习使用 K 均值算法对数据集的数据点进行聚类或分组处理。

操作步骤

1. 我们将使用 cpu 数据集通过简单的 K 均值算法对它的数据点进行聚类处理。在 Weka 安装目录的 data 目录下，你会找到 cpu 数据集。

   就像前面一节所做的那样，我们会用到两个实例变量，第一个变量用来保存 cpu 数据集的数据点，第二个变量是 SimpleKmeans 聚类器。

```
Instances cpu = null;
SimpleKMeans kmeans;
```

2. 然后，创建 loadArff 方法，用来加载并读取 cpu 数据集的内容。请注意，由于聚类是一种非监督方法，所以我们不必为数据集指定分类属性。

```
public void loadArff(String arffInput){
  DataSource source = null;
  try {
    source = new DataSource(arffInput);
    cpu = source.getDataSet();
  } catch (Exception e1) {
  }
}
```

3. 接下来，创建 clusterData 方法来建立聚类器。

```
public void clusterData(){
```

4. 把聚类器实例化，并且把种子值设置为 10。种子用来产生随机数，它是一个整型值。

```
kmeans = new SimpleKMeans();
kmeans.setSeed(10)
```

5. 然后，让聚类器保持数据实例的原始顺序。如果你觉得不需要保持数据集中实例原始的顺序，只要把 setPreserveInstancesOrder() 方法的参数设为 false 即可。然后把簇数设置为 10，最后使用 cpu 数据集创建聚类器。

```
try {
  kmeans.setPreserveInstancesOrder(true);
  kmeans.setNumClusters(10);
  kmeans.buildClusterer(cpu);
```

6. 接着，使用 for 循环，获取每个实例，以及简单地将 K 均值算法指派给它们的簇编号。

```
int[] assignments = kmeans.getAssignments();
  int i = 0;
  for(int clusterNum : assignments) {
```

```
            System.out.printf("Instance %d -> Cluster %d\n", i,
               clusterNum);
            i++;
         }
      } catch (Exception e1) {
      }
```

示例的完整代码如下:

```
import weka.clusterers.SimpleKMeans;
import weka.core.Instances;
import weka.core.converters.ConverterUtils.DataSource;

public class WekaClusterTest {
   Instances cpu = null;
   SimpleKMeans kmeans;

   public void loadArff(String arffInput){
      DataSource source = null;
      try {
         source = new DataSource(arffInput);
         cpu = source.getDataSet();
      } catch (Exception e1) {
      }
   }

   public void clusterData(){
      kmeans = new SimpleKMeans();
      kmeans.setSeed(10);
      try {
         kmeans.setPreserveInstancesOrder(true);
         kmeans.setNumClusters(10);
         kmeans.buildClusterer(cpu);
         int[] assignments = kmeans.getAssignments();
         int i = 0;
         for(int clusterNum : assignments) {
            System.out.printf("Instance %d -> Cluster %d\n", i,
               clusterNum);
            i++;
         }
      } catch (Exception e1) {
      }
   }
```

```
    public static void main(String[] args) throws Exception{
        WekaClusterTest test = new WekaClusterTest();
        test.loadArff("path to cpu.arff file");
        test.clusterData();
    }
}
```

`cpu.arff` 文件拥有 209 个数据实例，前 10 个输出如下：

```
Instance 0 -> Cluster 7
Instance 1 -> Cluster 5
Instance 2 -> Cluster 5
Instance 3 -> Cluster 5
Instance 4 -> Cluster 1
Instance 5 -> Cluster 5
Instance 6 -> Cluster 5
Instance 7 -> Cluster 5
Instance 8 -> Cluster 4
Instance 9 -> Cluster 4
```

4.9 依据类别对数据进行聚类处理

对于那些带有类别信息的数据集（对于非监督学习而言，这种情形并不常见），Weka 提供了依据类别进行聚类的方法。在这个方法中，Weka 在进行聚类处理时先忽略分类属性，然后在测试阶段，根据每个簇分类属性的多数值把分类指派给各个簇。本部分，我们将讲解这种方法。

操作方法

1. 本部分，我们使用的是带有分类值的数据集，即 `weather.nominal.arff` 文件，你在 Weka 的 data 目录中可以找到它。

   代码中，我们创建两个实例变量，第一个变量用来保存数据集实例，第二个变量是一个 EM（期望最小化）聚类器。

   ```
   Instances weather = null;
   EM clusterer;
   ```

2. 接下来，创建 loadArff 方法，用来加载并读取数据集，把最后一个索引设为它的分类索引。

    ```
    public void loadArff(String arffInput){
      DataSource source = null;
      try {
        source = new DataSource(arffInput);
        weather = source.getDataSet();
        weather.setClassIndex(weather.numAttributes() - 1);
      } catch (Exception e1) {
      }
    }
    ```

3. 然后，创建本部分最重要的方法，即用来根据分类生成簇集的方法。

    ```
    public void generateClassToCluster(){
    ```

4. 为此，我们先创建一个 Remove 过滤器，用来从数据集中移除分类属性，在进行聚类处理时，Weka 会忽略这个属性。

    ```
    Remove filter = new Remove();
    filter.setAttributeIndices("" + (weather.classIndex() + 1));
    ```

5. 然后，把过滤器应用到数据集上。

    ```
    try {
      filter.setInputFormat(weather);
    ```

6. 我们将获取不带类别变量的数据集，并且依据数据创建一个 EM（期望最小化）聚类器。

    ```
    Instances dataClusterer = Filter.useFilter(weather, filter);
    clusterer = new EM();
    clusterer.buildClusterer(dataClusterer);
    ```

7. 然后，使用原数据集的分类信息对聚类进行评估。

    ```
    ClusterEvaluation eval = new ClusterEvaluation();
    eval.setClusterer(clusterer);
    eval.evaluateClusterer(weather);
    ```

8. 最后，在控制台中把聚类结果打印出来。

```
      System.out.println(eval.clusterResultsToString());
    } catch (Exception e) {
    }
}
```

示例的完整代码整理如下：

```
import weka.clusterers.ClusterEvaluation;
import weka.clusterers.EM;
import weka.core.Instances;
import weka.core.converters.ConverterUtils.DataSource;
import weka.filters.Filter;
import weka.filters.unsupervised.attribute.Remove;

public class WekaClassesToClusterTest {
    Instances weather = null;
    EM clusterer;

    public void loadArff(String arffInput){
        DataSource source = null;
        try {
            source = new DataSource(arffInput);
            weather = source.getDataSet();
            weather.setClassIndex(weather.numAttributes() - 1);
        } catch (Exception e1) {
        }
    }

    public void generateClassToCluster(){
        Remove filter = new Remove();
        filter.setAttributeIndices("" + (weather.classIndex() + 1));
        try {
            filter.setInputFormat(weather);
            Instances dataClusterer = Filter.useFilter(weather, filter);
            clusterer = new EM();
            clusterer.buildClusterer(dataClusterer);
            ClusterEvaluation eval = new ClusterEvaluation();
            eval.setClusterer(clusterer);
            eval.evaluateClusterer(weather);

            System.out.println(eval.clusterResultsToString());
```

```
        } catch (Exception e) {
        }
    }
    public static void main(String[] args){
        WekaClassesToClusterTest test = new WekaClassesToClusterTest();
        test.loadArff("path to weather.nominal.arff file");
        test.generateClassToCluster();
    }
}
```

4.10　学习数据间的关联规则

关联规则学习也是一种机器学习技术，它用来在数据集的不同特征与变量之间发现关联与规则。统计中也有类似的技术叫相关性，相关内容在第 3 章中已经讲解过。但是在进行决策时，关联规则学习更加有用。比如，通过分析超市大数据，机器学习器会发现：如果一个顾客购买了洋葱、西红柿、鸡肉饼、蛋黄酱，那么她很有可能还会买圆面包来做汉堡包。

本部分，我们将学习如何使用 Weka 从数据集学习关联规则。

准备工作

本部分学习中，我们会用到超市数据集，你在 Weka 的 `data` 目录中可以找到它。这个数据集总共包含 4 627 个实例，每个实例都带有 217 个二元属性。这些属性的值要么是 `true` 要么是 `missing`。其中有一个叫 `total` 的名义类属性，当交易额低于$100 时，其值为 `low`；当交易额高于$100 时，其值为 `high`。

操作步骤

1. 首先声明两个实例变量，第一个用来保存超市数据集的数据，第二个表示 Apriori 学习器。

    ```
    Instances superMarket = null;
    Apriori apriori;
    ```

2. 而后创建 loadArff()方法，用来加载与读取数据集。本部分，我们不必为数据集设置分类属性。

```
public void loadArff(String arffInput){
  DataSource source = null;
  try {
    source = new DataSource(arffInput);
    superMarket = source.getDataSet();
  } catch (Exception e1) {
  }
}
```

3. 创建 generateRule()方法，先把 Apriori 学习器实例化，而后使用给定的数据集创建关联关系。最后，在控制台中把学习器显示出来。

```
public void generateRule(){
  apriori = new Apriori();
    try {
      apriori.buildAssociations(superMarket);
      System.out.println(apriori);
    } catch (Exception e) {
    }
}
```

Apriori 学习器生成的规则数默认为 10。如果需要生成更多规则，请在创建关联关系之前，调用 setNumRules(n)方法设置规则数，其中参数 n 是一个整数，表示规则数目，如下所示：`learn-apriori.setNumRules(n)`。

示例的完整代码整理如下：

```
import weka.associations.Apriori;
import weka.core.Instances;
import weka.core.converters.ConverterUtils.DataSource;

public class WekaAssociationRuleTest {
  Instances superMarket = null;
  Apriori apriori;
  public void loadArff(String arffInput){
    DataSource source = null;
    try {
      source = new DataSource(arffInput);
      superMarket = source.getDataSet();
    } catch (Exception e1) {
    }
  }
```

```
public void generateRule(){
    apriori = new Apriori();
    try {
       apriori.buildAssociations(superMarket);
       System.out.println(apriori);
    } catch (Exception e) {
    }
}
public static void main(String args[]){
    WekaAssociationRuleTest test = new WekaAssociationRuleTest();
    test.loadArff("path to supermarket.arff file");
    test.generateRule();
}
```

运行上面代码，Apriori 学习器所找到的规则如下：

```
1. biscuits=t frozen foods=t fruit=t total=high 788 ==> bread and cake=t
723    <conf:(0.92)> lift:(1.27) lev:(0.03) [155] conv:(3.35)
2. baking needs=t biscuits=t fruit=t total=high 760 ==> bread and cake=t
696    <conf:(0.92)> lift:(1.27) lev:(0.03) [149] conv:(3.28)
3. baking needs=t frozen foods=t fruit=t total=high 770 ==> bread and
cake=t 705 <conf:(0.92)> lift:(1.27) lev:(0.03) [150] conv:(3.27)
4. biscuits=t fruit=t vegetables=t total=high 815 ==> bread and cake=t 746
<conf:(0.92)> lift:(1.27) lev:(0.03) [159] conv:(3.26)
5. party snack foods=t fruit=t total=high 854 ==> bread and cake=t 779
<conf:(0.91)> lift:(1.27) lev:(0.04) [164] conv:(3.15)
6. biscuits=t frozen foods=t vegetables=t total=high 797 ==> bread and
cake=t 725 <conf:(0.91)> lift:(1.26) lev:(0.03) [151] conv:(3.06)
7. baking needs=t biscuits=t vegetables=t total=high 772 ==> bread and
cake=t 701 <conf:(0.91)> lift:(1.26) lev:(0.03) [145] conv:(3.01)
8. biscuits=t fruit=t total=high 954 ==> bread and cake=t 866
<conf:(0.91)> lift:(1.26) lev:(0.04) [179] conv:(3)
9. frozen foods=t fruit=t vegetables=t total=high 834 ==> bread and cake=t
757 <conf:(0.91)> lift:(1.26) lev:(0.03) [156] conv:(3)
10. frozen foods=t fruit=t total=high 969 ==> bread and cake=t 877
<conf:(0.91)> lift:(1.26) lev:(0.04) [179] conv:(2.92)
```

4.11 使用低层方法、过滤方法、元分类器方法选择特征/属性

特征选择是机器学习中一个重要的处理过程，用来从数据集的一系列属性中识别出最重要的属性。基于这些选取的属性生成分类器，与带有属性的分类器相比，这种分类器能

够产生更好的分类结果。

针对属性选择，Weka 提供了 3 种方法，分别为低层属性选择方法、使用过滤器选择属性、使用元分类器选择属性。本部分将讲解 Weka 中这 3 种属性选择技术。

准备工作

在讲解属性选择技术的过程中，我们将用到 iris 数据集，在 Weka 安装目录的 data 目录中你可以找到它。

进行属性选择时需要两个必备元素，即搜索方法与评估方法。本部分，我们将使用最佳优先搜索（Best First Search）作为搜索方法，使用基于关联的特征子集选择（Correlation-based Feature Subset Selection）作为子集评估方法。

操作步骤

1. 首先声明一个实例变量，用来保存 iris 数据集中的数据。而后声明另一个变量作为 NaiveBayes 分类器。

   ```
   Instances iris = null;
   NaiveBayes nb;
   ```

2. 创建 loadArff() 方法，加载数据集。这个方法还用来读取数据实例，把数据集的最后一个属性设置为类别属性。

   ```
   public void loadArff(String arffInput){
     DataSource source = null;
     try {
       source = new DataSource(arffInput);
       iris = source.getDataSet();
       iris.setClassIndex(iris.numAttributes() - 1);
     } catch (Exception e1) {
     }
   }
   ```

3. 我们先从简单的开始，即先创建一个方法，它使用 Weka 的低层属性选择方法。

   ```
   public void selectFeatures(){
   ```

4. 创建一个 AttributeSelection 对象。

```
AttributeSelection attSelection = new AttributeSelection();
```

5. 接着,创建搜索与评估器对象,并把它们的设置给属性选择对象。

```
CfsSubsetEval eval = new CfsSubsetEval();
BestFirst search = new BestFirst();
attSelection.setEvaluator(eval);
attSelection.setSearch(search);
```

6. 然后,使用属性选择对象,借助搜索与评估器从 iris 数据集选择属性。我们会得到所选属性的索引,并把所选属性的编号(属性编号从 0 开始)显示出来。

```
try {
  attSelection.SelectAttributes(iris);
  int[] attIndex = attSelection.selectedAttributes();
  System.out.println(Utils.arrayToString(attIndex));
} catch (Exception e) {
}
}
```

这个方法的输出如下:

```
2,3,4
```

输出表明属性选择技术从 iris 数据集的所有属性中选择了编号为 2、3、4 的属性。

7. 接下来,我们要创建一个方法,实现另一种属性选择技术,即基于过滤器的属性选择技术。

```
public void selectFeaturesWithFilter(){
```

8. 创建一个属性选择过滤器。请注意,这个过滤器的包不是我们在本部分第一个方法中所使用的那个。

```
weka.filters.supervised.attribute.AttributeSelection filter =
  new weka.filters.supervised.attribute.AttributeSelection();
```

9. 接着,创建搜索与评估器对象,并为过滤器设置评估器与搜索对象。

```
CfsSubsetEval eval = new CfsSubsetEval();
BestFirst search = new BestFirst();
filter.setEvaluator(eval);
```

4.11 使用低层方法、过滤方法、元分类器方法选择特征/属性

```
filter.setSearch(search);
```

10. 然后，把过滤器应用到 iris 数据集，调用 Filter 类的 useFilter() 方法来取回新数据，这个方法有两个参数，分别用来接收数据集与过滤器。这与我们在前面方法中所看到的有些不同。如果想在运行中通过选择属性（由过滤技术选出）来创建一个新的 ARFF 文件，这是非常有用的。

    ```
    try {
      filter.setInputFormat(iris);
      Instances newData = Filter.useFilter(iris, filter);
      System.out.println(newData);
    } catch (Exception e) {
    }
    }
    ```

    从控制台中的输出，我们可以看到在 ARFF 文件的属性部分有如下内容：

    ```
    @attribute petallength numeric
    @attribute petalwidth numeric
    @attribute class {Iris-setosa,Iris-versicolor,Iris-virginica}
    ```

    这表示有两个属性与一个类别属性被我们刚刚使用的属性选择方法选出。

11. 最后，创建 selectFeaturesWithClassifiers() 方法，用来在把数据集传递给分类器（本示例中是 NaiveBayes 分类器）之前将属性选出来。

    ```
    public void selectFeaturesWithClassifiers(){
    ```

12. 创建一个元分类器，用来在把数据传递给 NaiveBayes 分类器之前，减少数据的维数（即选择属性）。

    ```
    AttributeSelectedClassifier classifier = new
      AttributeSelectedClassifier();
    ```

13. 创建一个评估器、一个搜索对象与一个 NaiveBayes 分类器。

    ```
    CfsSubsetEval eval = new CfsSubsetEval();
    BestFirst search = new BestFirst();
    nb = new NaiveBayes();
    ```

14. 把评估器、搜索对象、NaiveBayes 分类器设置给元分类器。

```
classifier.setClassifier(nb);
classifier.setEvaluator(eval);
classifier.setSearch(search);
```

15. 接下来，我们将使用元分类器技术选择的属性来评估 NaiveBayes 分类器的性能。请注意，本示例中元分类器选择的属性像个黑盒。做评估时，我们将使用 10 折交叉验证方法。

```
Evaluation evaluation;
try {
    evaluation = new Evaluation(iris);
    evaluation.crossValidateModel(classifier, iris, 10, new
      Random(1));
    System.out.println(evaluation.toSummaryString());
} catch (Exception e) {
}
}
```

示例完整代码整理如下：

```
import java.util.Random;
import weka.attributeSelection.AttributeSelection;
import weka.attributeSelection.BestFirst;
import weka.attributeSelection.CfsSubsetEval;
import weka.classifiers.Evaluation;
import weka.classifiers.bayes.NaiveBayes;
import weka.classifiers.meta.AttributeSelectedClassifier;
import weka.core.Instances;
import weka.core.Utils;
import weka.core.converters.ConverterUtils.DataSource;
import weka.filters.Filter;

public class WekaFeatureSelectionTest {
    Instances iris = null;
    NaiveBayes nb;
    public void loadArff(String arffInput){
        DataSource source = null;
        try {
            source = new DataSource(arffInput);
            iris = source.getDataSet();
            iris.setClassIndex(iris.numAttributes() - 1);
        } catch (Exception e1) {
```

```java
    }
}

public void selectFeatures(){
    AttributeSelection attSelection = new AttributeSelection();
    CfsSubsetEval eval = new CfsSubsetEval();
    BestFirst search = new BestFirst();
    attSelection.setEvaluator(eval);
    attSelection.setSearch(search);
    try {
        attSelection.SelectAttributes(iris);
        int[] attIndex = attSelection.selectedAttributes();
        System.out.println(Utils.arrayToString(attIndex));
    } catch (Exception e) {
    }
}

public void selectFeaturesWithFilter(){
    weka.filters.supervised.attribute.AttributeSelection filter = new
        weka.filters.supervised.attribute.AttributeSelection();
    CfsSubsetEval eval = new CfsSubsetEval();
    BestFirst search = new BestFirst();
    filter.setEvaluator(eval);
    filter.setSearch(search);
    try {
        filter.setInputFormat(iris);
        Instances newData = Filter.useFilter(iris, filter);
        System.out.println(newData);
    } catch (Exception e) {
    }
}
public void selectFeaturesWithClassifiers(){
    AttributeSelectedClassifier classifier = new
        AttributeSelectedClassifier();
    CfsSubsetEval eval = new CfsSubsetEval();
    BestFirst search = new BestFirst();
    nb = new NaiveBayes();
    classifier.setClassifier(nb);
    classifier.setEvaluator(eval);
    classifier.setSearch(search);
    Evaluation evaluation;
    try {
        evaluation = new Evaluation(iris);
```

```
            evaluation.crossValidateModel(classifier, iris, 10, new
               Random(1));
            System.out.println(evaluation.toSummaryString());
        } catch (Exception e) {
        }
    }
    public static void main(String[] args){
        WekaFeatureSelectionTest test = new WekaFeatureSelectionTest();
        test.loadArff("C:/Program Files/Weka-3-6/data/iris.arff");
        test.selectFeatures();
        test.selectFeaturesWithFilter();
        test.selectFeaturesWithClassifiers();
    }
```

运行上面代码,得到如下输出结果:

```
Correctly Classified Instances          144               96      %
Incorrectly Classified Instances          6                4      %
Kappa statistic                         0.94
Mean absolute error                     0.0286
Root mean squared error                 0.1386
Relative absolute error                 6.4429 %
Root relative squared error            29.4066 %
Total Number of Instances              150
```

 我们可以使用 Weka(包括 API 及其 GUI)进行各种不同的机器学习任务,网上也有相关的视频教程。

第 5 章
数据学习 II

本章涵盖如下内容：

- 使用 Java 机器学习库向数据应用机器学习；
- 导入与导出数据集；
- 聚类与评估；
- 分类；
- 交叉验证与 held-out 测试；
- 特征计分；
- 特征选择；
- 使用斯坦福分类器对数据点分类；
- 使用 MOA（Massive Online Analysis）对数据点分类。

5.1 简介

在第 4 章，我们学习了如何使用 Weka 机器学习平台来进行分类、聚类、关联规则挖掘、特征选择等。我们还提到 Weka 不是唯一一个使用 Java 语言用来学习数据模式的工具。除了 Weka 之外，还有许多其他工具可以用来执行类似的任务，比如 Java-ML、MOA、斯坦福机器学习库等。

本章，我们将重点学习这些工具中的几个来对数据进行机器学习分析。

5.2 使用 Java 机器学习库（Java-ML）向数据应用机器学习

Java 机器学习库（Java-ML）包含一系列标准的机器学习算法。与 Weka 不同，这个库不带有任何 GUI 界面，因为它主要面向的是软件开发人员。Java-ML 特别好的地方是每种算法都有相同接口，因此算法实现起来相当容易、简单易懂。另一个优点是这个库拥有丰富的支持，源代码文档齐全，并且容易扩展，有大量使用这个库实现的机器学习任务的代码示例与教程。关于这个库的更多细节，请参阅 java-ML 官网。

本部分，我们将使用这个库做如下任务：

- 导入与导出数据集；
- 聚类与评估；
- 分类；
- 交叉验证与 held-out 测试；
- 特征计分；
- 特征选择。

准备工作

开始之前，先做如下准备。

1. 如图 5-1 所示，在本部分，我们会使用 Java-ML 0.1.7 版本。请读者自行下载。

图 5-1

2. 下载完成后，对 zip 文件解压，得到如图 5-2 所示的目录。

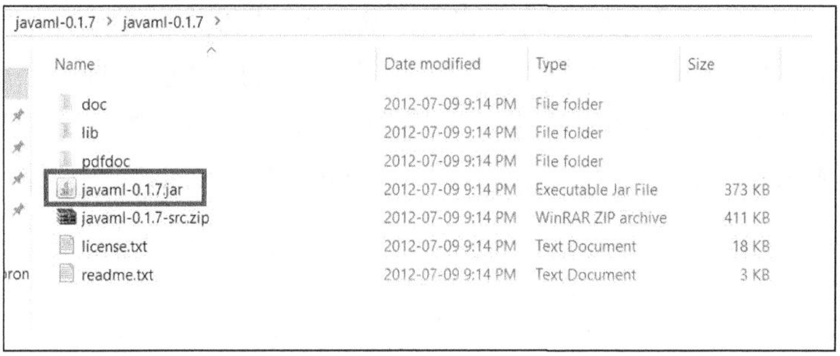

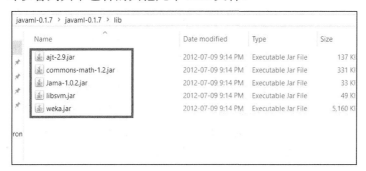

图 5-2

把 `javaml-0.1.7.jar` 文件作为外部文件添加到你用来实现本部分内容的 Eclipse 项目中。

3. 在解压后的目录中，还包含一个名为 `lib` 的文件夹。如图 5-3 所示，打开 lib 文件夹后，可以看到其中包含的其他几个 JAR 文件。

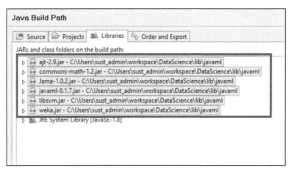

图 5-3

如图 5-4 所示，由于 Java-ML 需要这些 JAR 文件，因此，我们还要把这些 JAR 文件作为外部文件添加到项目中。

图 5-4

4. 本部分，我们还会用到 Iris 数据集，它与 Java-ML 原生文件格式相兼容。但是 Iris 与其他数据文件并不包含在这个库中，需要从不同的地方进行下载。请下载这些数据集。Java-ML 支持两种类型的数据集，分别是 111 个小型 UCI 数据集与 7 个大型 UCI 数据集。为了方便进行练习，强烈建议你把这两种数据集全都下载下来。如图 5-5 所示，我们单击 "Download 111 small UCI datasets"，根据提示下载即可。

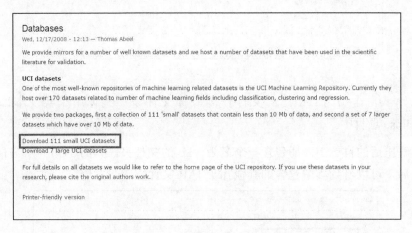

图 5-5

5. 下载完成后，进行解压缩，你将看到 111 个文件夹，每个文件夹包含一个数据集。找到 iris 数据集所在的文件夹，打开它，你将看到两个数据文件与一个名称文件。如图 5-6 所示，在本部分，我们将使用 `iris.data` 文件。你需要把这个文件的路径记下来，后面学习会用到这个路径。

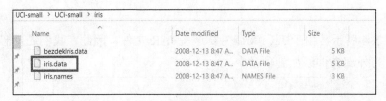

图 5-6

 在使用 UCI 数据集时，你需要给出原始引用地址。

操作步骤

1. 创建一个名为 `JavaMachineLearning` 的类。我们将使用 main 方法来实现所有

5.2 使用 Java 机器学习库（Java-ML）向数据应用机器学习

机器学习任务，它会抛出 `IOException` 异常。

```
public class JavaMachineLearning {
    public static void main(String[] args) throws IOException{
```

2. 首先，调用 Java-ML 库中的 `FileHandler` 类的 `loadDataset()` 方法来读取 iris 数据集。

```
Dataset data = FileHandler.loadDataset(new File("path to your
   iris.data"), 4, ",");
```

这个方法的参数分别为数据集的路径、分类属性的位置，以及值分隔符。我们可以使用任意一个标准的文本编辑器来读取数据集。属性的起始索引为 0，iris 数据集的类属性是它的第五个属性。因此，把第二个参数设置为 1。并且本例中，我们将使用逗号来分隔数值。所以，把第三个参数设置为逗号。读取文件的内容后，将其放入一个数据集对象中。

3. 通过把数据集对象传递给 `System.out.println()` 方法，即可将数据集的内容打印出来。

```
System.out.println(data);
```

4. 代码的部分输出如下所示：

```
[{[5.1, 3.5, 1.4, 0.2];Iris-setosa}, {[4.9, 3.0, 1.4,
   0.2];Iris-setosa}, {[4.7, 3.2, 1.3, 0.2];Iris-setosa}, {[4.6,
      3.1, 1.5, 0.2];Iris-setosa}, {[5.0, 3.6, 1.4, 0.2];Irissetosa},
   {[5.4, 3.9, 1.7, 0.4];Iris-setosa}, {[4.6, 3.4,
      1.4, 0.3];Iris-setosa}, {[5.0, 3.4, 1.5, 0.2];Irissetosa},
         {[4.4, 2.9, 1.4, 0.2];Iris-setosa}, ...]
```

5. 有时我们需要把数据集从 .data 格式导出为 .txt 格式，为此 Java-ML 提供了一个非常简单的方法，即调用 `FileHandler` 类的 `exportDataset()` 方法。这个方法带有两个参数，第一个参数是数据，第二个参数是输出文件。下面这行代码用来在 C:/ 目录下创建一个文本文件，里面包含着 iris 数据集的内容。

```
FileHandler.exportDataset(data, new File("c:/javamloutput.
   txt"));
```

前面代码所生成的文本文件的部分输出如下所示：

```
Iris-setosa  5.1  3.5  1.4  0.2
Iris-setosa  4.9  3.0  1.4  0.2
Iris-setosa  4.7  3.2  1.3  0.2
Iris-setosa  4.6  3.1  1.5  0.2
..................................
```

> 对于 Java-ML 生成的数据文件，有两个地方需要注意：首先，类值是第一个属性；其次，分隔各个值时不再使用逗号（在原来的 .data 文件中采用逗号分隔），使用的是制表符。

6. 为了读取上一步中创建的数据文件，再次调用 `loadDataset()` 方法，但是传入的参数值有所不同。

   ```
   data = FileHandler.loadDataset(new File("c:/javamloutput.
       txt"), 0,"\t");
   ```

7. 使用如下代码，打印数据：

   ```
   System.out.println(data);
   ```

   所得到的输出结果与步骤 3 中看到的输出结果一样。

   ```
   [{[5.1, 3.5, 1.4, 0.2];Iris-setosa}, {[4.9, 3.0, 1.4,
      0.2];Iris-setosa}, {[4.7, 3.2, 1.3, 0.2];Iris-setosa}, {[4.6,
       3.1, 1.5, 0.2];Iris-setosa}, {[5.0, 3.6, 1.4, 0.2];Irissetosa},
         {[5.4, 3.9, 1.7, 0.4];Iris-setosa}, {[4.6, 3.4,
          1.4, 0.3];Iris-setosa}, {[5.0, 3.4, 1.5, 0.2];Irissetosa},
            {[4.4, 2.9, 1.4, 0.2];Iris-setosa}, ...]
   ```

8. Java-ML 提供了非常简单的接口，用来进行聚类、显示簇集，以及评估聚类结果。本部分，我们将使用 KMeans 聚类。首先创建一个 KMeans 聚类器。

   ```
   Clusterer km = new KMeans();
   ```

9. 调用 `cluster()` 方法，把数据提供给聚类器。这样所得到的结果是数据点（或多个数据集）的多个簇集。把结果放入一个 Dataset 数组中。

   ```
   Dataset[] clusters = km.cluster(data);
   ```

10. 如果你想查看每个簇集中的数据点，使用一个 `for` 循环在数据集数组上进行迭代即可。

```
for(Dataset cluster:clusters){
  System.out.println("Cluster: " + cluster);
}
```

上面代码的部分输出结果如下：

```
Cluster: [{[6.3, 3.3, 6.0, 2.5];Iris-virginica}, {[7.1, 3.0,
  5.9, 2.1];Iris-virginica}, ...]
Cluster: [{[5.5, 2.3, 4.0, 1.3];Iris-versicolor}, {[5.7, 2.8,
  4.5, 1.3];Iris-versicolor}, ...]
Cluster: [{[5.1, 3.5, 1.4, 0.2];Iris-setosa}, {[4.9, 3.0, 1.4,
  0.2];Iris-setosa}, ...]
Cluster: [{[7.0, 3.2, 4.7, 1.4];Iris-versicolor}, {[6.4, 3.2,
  4.5, 1.5];Iris-versicolor}, ...]
```

从上面输出中，我们可以看到 KMeans 算法从 iris 数据集创建了 4 个簇集。

11. 误差平方和是用来评估聚类器性能的指标之一。我们将使用 `ClusterEvaluation` 类来测量聚类误差。

    ```
    ClusterEvaluation sse = new SumOfSquaredErrors();
    ```

12. 接下来，我们把簇集传递给误差平方和对象的计分方法，计算聚类的误差平方和。

    ```
    double score = sse.score(clusters);
    ```

13. 打印误差分数。

    ```
    System.out.println(score);
    ```

    输出结果如下：

    ```
    114.9465465309897
    ```

    上面这个值就是对 iris 数据集进行 KMeans 聚类的误差平方和。

14. 使用 Java-ML 进行分类也非常简单，只需要几行代码即可搞定。下面的代码用来创建一个 KNN 分类器（K 最近邻分类器）。这个分类器使用最靠近的 5 个邻居的多数投票来预测新数据点的标签。`buildClassifier()` 方法用来训练一个分类器，其参数是一个数据集（本例中是 iris 数据集）。

```
Classifier knn = new KNearestNeighbors(5);
knn.buildClassifier(data);
```

15. 模型创建好之后，接下来就要对模型进行评估。这里我们将学习使用 Java-ML 实现两种评估方法。

    - K 折交叉验证
    - Held-out 测试

16. 对于 KNN 分类器的 K 折交叉验证，我们将使用分类器创建一个 `CrossValidation` 实例。`CrossValidation` 类有一个名为 `crossValidation()` 的方法，其参数为数据集。这个方法会返回一个 Map，第一个参数是对象，第二个参数是评估指标。

    ```
    CrossValidation cv = new CrossValidation(knn);
    Map<Object, PerformanceMeasure> cvEvaluation =
        cv.crossValidation(data);
    ```

17. 在得到交叉验证结果之后，我们可以使用如下语句把它们简单地打印出来。

    ```
    System.out.println(cvEvaluation);
    ```

    这会把每个分类的真正、假正、真负与假负显示出来。

    ```
    {Iris-versicolor=[TP=47.0, FP=1.0, TN=99.0, FN=3.0], Irisvirginica=[
      TP=49.0, FP=3.0, TN=97.0, FN=1.0], Iris-setosa=
         [TP=50.0, FP=0.0, TN=100.0, FN=0.0]}
    ```

18. 为了进行 held-out 测试，我们需要有一个测试数据集。但是不幸的是，我们没有任何 iris 的测试数据集。因此，我们把同一个 iris.data 文件（它用来训练 KNN 分类器）作为我们的测试数据集。但是，请注意，在真实情况下，你会有一个测试数据集，其属性数目与训练数据集中的属性数目完全一样，而数据点的标签则是未知的。

    首先，加载测试数据集。

    ```
    Dataset testData = FileHandler.loadDataset(new File("path to
        your iris.data "), 4, ",");
    ```

    然后，使用如下代码获取分类器在测试数据上的性能。

```
Map<Object, PerformanceMeasure> testEvaluation =
  EvaluateDataset.testDataset(knn, testData);
```

19. 接下来，通过遍历 Map 对象，为每个类打印结果。

    ```
    for(Object classVariable:testEvaluation.keySet()){
      System.out.println(classVariable + " class has " +
        testEvaluation.get(classVariable).getAccuracy());
    }
    ```

    上面代码将为每个分类打印 KNN 分类器的准确度。

    ```
    Iris-versicolor class has 0.9666666666666667
    Iris-virginica class has 0.9666666666666667
    Iris-setosa class has 1.0
    ```

20. 特征打分（Feature Scoring）是机器学习中用来减少维数的一个关键技术。在 Java-ML 中，我们将实现如下方法，用来为给定的属性计算分数。

    ```
    public double score(int attIndex);
    ```

    首先，创建一个特征打分算法实例。本部分，我们将使用信息增益率算法。

    ```
    GainRatio gainRatio = new GainRatio();
    ```

21. 接着，把算法应用到数据。

    ```
    gainRatio.build(data);
    ```

22. 最后，使用 `for` 循环，把属性索引逐个传递给 `score()` 方法，将每个特征的分数打印出来。

    ```
    for (int i = 0; i < gainRatio.noAttributes(); i++){
      System.out.println(gainRatio.score(i));
    }
    ```

    iris 数据集的特征分数如下：

    ```
    0.2560110727706682
    0.1497001925156687
    0.508659832906763
    0.4861382158327255
    ```

23. 此外，我们还可以依据一些特征排序算法对特征进行排序。为此，我们将实现 Java-ML 的 rank() 方法，其工作方式类似于 score() 方法，它们都以属性索引作为参数。

    ```
    public int rank(int attIndex);
    ```

    创建一个特征排序算法实例。本例中，我们将采用特征的 SVM 排序，它基于特征的递归消除法实现。构造函数的参数是最差排序特征的百分率，这些特征是要删除的。

    ```
    RecursiveFeatureEliminationSVM featureRank = new
        RecursiveFeatureEliminationSVM(0.2);
    ```

    - 接着，把算法应用到数据集。

    ```
    featureRank.build(data)
    ```

    - 最后，使用 for 循环，把属性索引依据次序传递给 rank()方法，把每个特征的排名打印出来。

    ```
    for (int i = 0; i < featureRank.noAttributes(); i++){
        System.out.println(featureRank.rank(i));
    }
    ```

    - iris 数据集的特征排序如下：

    ```
    3
    2
    0
    1
    ```

24. 在为了给特征打分与排序而获取个别特征信息过程中，应用 Java-ML 的特征子集选择时，我们只需要从那些来自于数据集的诸多特征中选取一个子集即可。

    首先，创建一个特征选择算法。本部分，我们将通过 greedy 方法使用正向选择方法。在选择特征的过程中，我们需要一个距离测度，这里我们选用皮尔逊相关系数。构造函数的第一个参数代表要在子集中选择的属性数目。

    ```
    GreedyForwardSelection featureSelection = new
        GreedyForwardSelection(5, new PearsonCorrelationCoefficient());
    ```

- 然后，把算法应用到数据集。

  ```
  featureSelection.build(data);
  ```

- 最后，使用如下语句把算法选择的特征打印出来。

  ```
  System.out.println(featureSelection.selectedAttributes());
  ```

输出的特征子集如下：

[0]

示例的完整代码整理如下：

```java
import java.io.File;
import java.io.IOException;
import java.util.Map;
import net.sf.javaml.classification.Classifier;
import net.sf.javaml.classification.KNearestNeighbors;
import net.sf.javaml.classification.evaluation.CrossValidation;
import net.sf.javaml.classification.evaluation.EvaluateDataset;
import net.sf.javaml.classification.evaluation.PerformanceMeasure;
import net.sf.javaml.clustering.Clusterer;
import net.sf.javaml.clustering.KMeans;
import net.sf.javaml.clustering.evaluation.ClusterEvaluation;
import net.sf.javaml.clustering.evaluation.SumOfSquaredErrors;
import net.sf.javaml.core.Dataset;
import net.sf.javaml.distance.PearsonCorrelationCoefficient;
import net.sf.javaml.featureselection.ranking.
   RecursiveFeatureEliminationSVM;
import net.sf.javaml.featureselection.scoring.GainRatio;
import net.sf.javaml.featureselection.subset.GreedyForwardSelection;
import net.sf.javaml.tools.data.FileHandler;

public class JavaMachineLearning {
  public static void main(String[] args) throws IOException{
    Dataset data = FileHandler.loadDataset(new File("path to
      iris.data"), 4, ",");
    System.out.println(data);
    FileHandler.exportDataset(data, new File("c:/javamloutput.
      txt"));
    data = FileHandler.loadDataset(new File("c:/javaml-output.txt"),
      0,"\t");
```

```java
System.out.println(data);
//聚类
Clusterer km = new KMeans();
Dataset[] clusters = km.cluster(data);
for(Dataset cluster:clusters){
    System.out.println("Cluster: " + cluster);
}
ClusterEvaluation sse= new SumOfSquaredErrors();
double score = sse.score(clusters);
System.out.println(score);
//分类
Classifier knn = new KNearestNeighbors(5);
knn.buildClassifier(data);
//交叉验证
CrossValidation cv = new CrossValidation(knn);
Map<Object, PerformanceMeasure> cvEvaluation =
  cv.crossValidation(data);
System.out.println(cvEvaluation);
//Held-out 测试
Dataset testData = FileHandler.loadDataset(new File("path to
   iris.data"), 4, ",");
Map<Object, PerformanceMeasure> testEvaluation =
      EvaluateDataset.testDataset(knn, testData);
for(Object classVariable:testEvaluation.keySet()){
    System.out.println(classVariable + " class has
      "+testEvaluation.get(classVariable).getAccuracy());
}
//特征打分
GainRatio gainRatio = new GainRatio();
gainRatio.build(data);
for (int i = 0; i < gainRatio.noAttributes(); i++){
    System.out.println(gainRatio.score(i));
}
//特征排序
RecursiveFeatureEliminationSVM featureRank = new
  RecursiveFeatureEliminationSVM(0.2);
featureRank.build(data);
for (int i = 0; i < featureRank.noAttributes(); i++){
    System.out.println(featureRank.rank(i));
}
```

```
        //特征子集选择
        GreedyForwardSelection featureSelection = new
          GreedyForwardSelection(5, new
              PearsonCorrelationCoefficient());
        featureSelection.build(data);
        System.out.println(featureSelection.selectedAttributes());
    }
}
```

5.3 使用斯坦福分类器对数据点分类

斯坦福分类器是一种机器学习分类器，由斯坦福大学的自然语言处理小组开发。它采用 Java 语言实现，使用最大熵（Maximum Entropy）作为它的分类器。最大熵类似于多类逻辑回归模型，但是参数的设置稍有不同。斯坦福分类器的优点是它采用了与 Google、Amazon 一样的基本技术。

准备工作

本部分，我们将使用斯坦福分类器借助于最大熵学习方法对数据点进行分类。这里我们选用 3.6.0 版本。更多细节请参阅：http://nlp.stanford.edu/software/classifier.html。为了运行本节代码，你需要安装 Java 8，并做如下准备。

1. 如图 5-7 所示，下载斯坦福分类器 3.6.0 版本。它在写作本书之时是最新版本，并且以 zip 压缩文件进行提供。

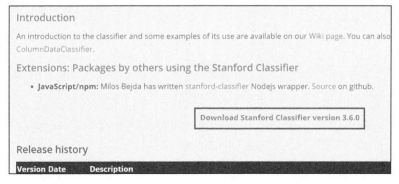

图 5-7

2. 下载完成后，解压缩文件，你将会看到如图 5-8 所示的文件与文件夹。

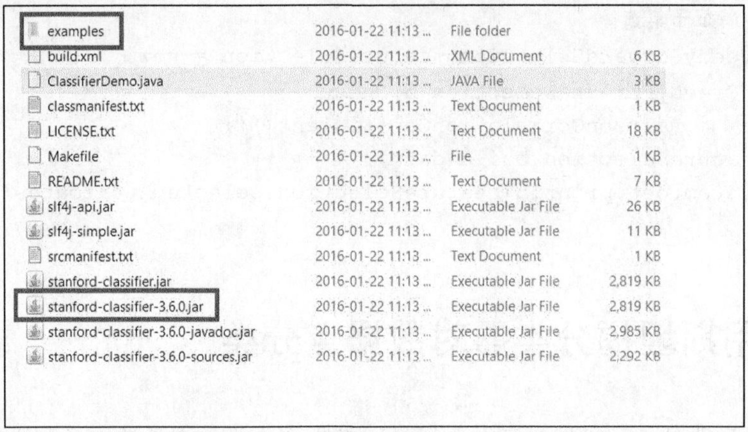

图 5-8

接着，如图 5-9 所示，我们需要把 stanford-classifier-3.6.0.jar 文件添加到 Eclipse 项目中。

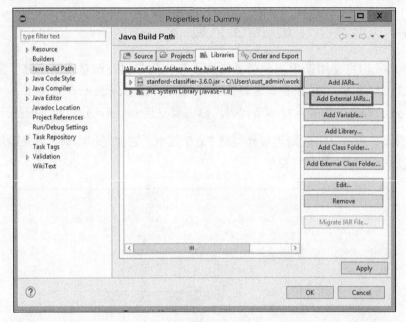

图 5-9

在解压缩后的文件夹中，还会看到一个名为 examples 的文件夹。本小节会用到这个文件夹中的内容。examples 文件夹包含两个数据集，分别是 cheeseDisease 数据集与 iris 数据集。每个数据集包含 3 个相关文件，分别是训练文件（扩展名为 .train）、测试文件（扩展名为 .test）与属性文件（扩展名为 .prop）。如图 5-10

所示，在本部分，我们会用到 cheeseDisease 数据集。

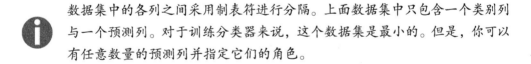

图 5-10

打开 cheeseDisease.train 文件，会看到如下内容：

```
2 Back Pain
2 Dissociative Disorders
2 Lipoma
1 Blue Rathgore
2 Gallstones
1 Chevrotin des Aravis
2 Pulmonary Embolism
2 Gastroenteritis
2 Ornithine Carbamoyltransferase Deficiency Disease ............
```

第一列（1 或 2）是数据实例的类别，第二列（字符串值）是名称。类别 1 后面紧跟着的是奶酪名称，类别 2 后面紧跟着的是疾病名称。在数据集上应用监督分类的目标是创建一个分类器，它可以把奶酪名称与疾病名称区分开来。

> 数据集中的各列之间采用制表符进行分隔。上面数据集中只包含一个类别列与一个预测列。对于训练分类器来说，这个数据集是最小的。但是，你可以有任意数量的预测列并指定它们的角色。

3. cheese2007.prop 文件是数据集的属性文件。你需要了解这个文件的内容，本部分代码需要从其中获取必要的类别特征信息、分类器所使用的特征类型、控制台中的显示格式，以及分类器的参数等。接下来，我们将查看这个文件的内容。

属性文件的头几行列出了特征项。其中以#开头的行是注释行。这些行会指示分类器在学习过程中使用类别特征（对应于训练文件中的第一列）。它还提供了其他一些信息，比如分类器要使用的 N-gram 特征，指出 N-gram 的最小长度是 1，最大长度是 4。而且，分类器还会在计算 N-gram 过程中使用前后缀，并指出分组长度（binnedLengths）为 10、20、30。

```
# # Features
# useClassFeature=true
1.useNGrams=true
1.usePrefixSuffixNGrams=true
1.maxNGramLeng=4
1.minNGramLeng=1
1.binnedLengths=10,20,30
```

接下来几行是 Mapping，其中，属性文件"告诉"分类器评估真实值是列 0，并且需要为预测值创建列 1。

```
# # Mapping
# goldAnswerColumn=0
displayedColumn=1
```

然后，属性文件保存最大熵分类器的优化参数。

```
#
# Optimization
# intern=true
 sigma=3
 useQN=true
 QNsize=15
 tolerance=1e-4
```

最后，属性文件给出了训练文件与测试文件的路径。

```
# Training input
# trainFile=./examples/cheeseDisease.train
testFile=./examples/cheeseDisease.test
```

操作步骤

1. 本部分，我们将在项目中创建一个类，并且只使用 main() 方法来进行分类演示。请注意，main() 方法可能会抛出异常。

```
public class StanfordClassifier { public static void
   main(String[] args) throws Exception {
```

2. 斯坦福分类器在 `ColumnDataClassifier` 中实现。创建一个分类器，并且把 cheeseDisease 数据集属性文件的路径提供给它。

   ```
   ColumnDataClassifier columnDataClassifier = new
      ColumnDataClassifier("examples/cheese2007.prop");
   ```

3. 接下来，使用训练数据，创建一个分类器。`Classifier` 类的泛型为`<String, String>`，第一列是类别，第二列是奶酪/疾病名称。请注意，虽然类别列值是 1 或 2，但我们仍然把它看作字符串。

   ```
   Classifier<String,String> classifier =
      columnDataClassifier.makeClassifier
        (columnDataClassifier.readTrainingExamples
          ("examples/cheeseDisease.train"));
   ```

4. 最后，遍历测试数据集的每一行。测试数据集与训练数据集类似，第一列是真实类别，第二列是名称。测试数据集的头几行如下：

   2 Psittacosis

   2 Cushing Syndrome

   2 Esotropia

   2 Jaundice, Neonatal

   2 Thymoma……………

工作原理

把列数据分类器应用到测试集的每一行，结果发送到一个 `Datum` 对象。分类器预测 `Datum` 对象的类别，把预测结果在控制台中打印出来。

```
for (String line :
ObjectBank.getLineIterator("examples/cheeseDisease.test", "utf-8")) {
Datum<String,String> d = columnDataClassifier.makeDatumFromLine(line);
System.out.println(line + " ==> " + classifier.classOf(d)); }
```

运行代码，在控制台中出现如下输出（以下截取了输出的一部分）：

2 Psittacosis ==> 2 2 Cushing Syndrome ==> 2 2 Esotropia ==> 2 2 Jaundice,

```
Neonatal ==> 2 2 Thymoma ==> 2 1 Caerphilly ==> 1 2 Teratoma ==> 2 2
Phantom Limb ==> 1 2 Iron Overload ==> 1 ...............
```

第一列是实际类别，第二列是名称，==>符号右侧的值是分类器预测得到的类别。

示例的完整代码整理如下：

```
import edu.stanford.nlp.classify.Classifier;
import edu.stanford.nlp.classify.ColumnDataClassifier;
import edu.stanford.nlp.ling.Datum;
import edu.stanford.nlp.objectbank.ObjectBank;
public class StanfordClassifier {
public static void main(String[] args) throws Exception {
ColumnDataClassifier columnDataClassifier = new
ColumnDataClassifier("examples/cheese2007.prop"); Classifier<String,String>
classifier =
columnDataClassifier.makeClassifier(columnDataClassifier.readTrainingExampl
es("examples/cheeseDisease.train"));
 for (String line :
ObjectBank.getLineIterator("examples/cheeseDisease.test", "utf-8")) {
Datum<String,String> d = columnDataClassifier.makeDatumFromLine(line);
System.out.println(line + " ==> " + classifier.classOf(d));
}
}
}
```

本部分未演示如何加载与保存斯坦福分类器模型。如果你感兴趣，请查看软件包中的 `ClassifierDemo.java` 文件。

5.4 使用 MOA 对数据点分类

MOA（Massive Online Analysis）是一个面向数据流挖掘的开源框架，它采用 Java 语言编写，关联了 Weka 项目，具有更好的扩展性。MOA 有着非常活跃的成长社区，包含一系列机器学习算法，比如分类、聚类、回归、概念漂移识别、推荐系统。MOA 最主要的优点是开发者可以很方便地扩展它，并且可以与 Weka 双向交互。

准备工作

开始之前，需要先做如下准备。

1. 下载 MOA，从 MOA 的启始页面（http://moa.cms.waikato.ac.nz/getting-started/）可以进行访问。

   如图 5-11 所示，把名为 moa-release-2016.04.zip 的 zip 文件下载到你的电脑中，保存到指定的目录下。

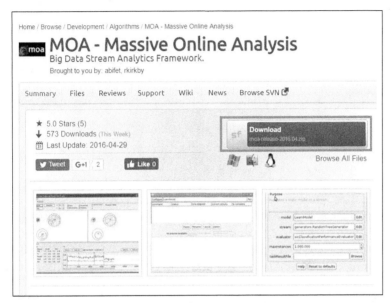

   图 5-11

2. 下载完成后，进行解压缩，你将看到如图 5-12 所示的文件与文件夹。

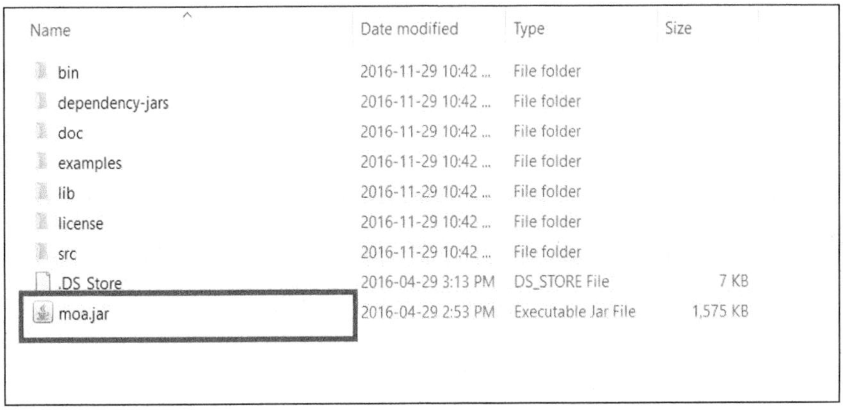

   图 5-12

3. 如图 5-13 所示，把 moa.jar 文件作为外部库添加到你的项目中。

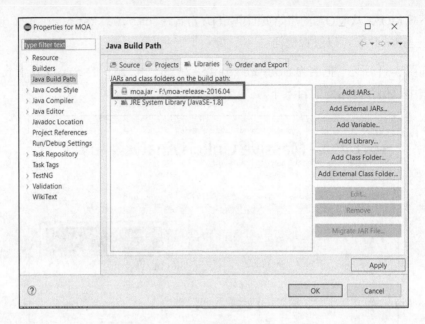

图 5-13

操作步骤

1. 首先，创建 MOA 类与 run()方法，该方法带有两个参数，第一个参数表示要处理的实例数量，第二个参数表示是否要对分类器进行测试。

   ```
   public class MOA { public void run(int numInstances, boolean
     isTesting){
   ```

2. 创建一个 HoeffdingTree 分类器。

   ```
   Classifier learner = new HoeffdingTree();
   ```

    MOA 实现了如下分类器：朴素贝叶斯、Hoeffding Tree、Hoeffding Option Tree、Hoeffding Adaptive Tree、Bagging、Boosting、Bagging using ADWIN、Leveraging Bagging、SGD、Perceptron、SPegasos。

3. 接下来，创建一个随机径向基函数流。
4. 准备要使用的流。

   ```
   stream.prepareForUse();
   ```

5. 设置引用到数据流的头部。使用 `getHeader()` 方法可以得到数据流头部。

   ```
   learner.setModelContext(stream.getHeader());
   ```

6. 然后，准备要使用的分类器。

   ```
   learner.prepareForUse();
   ```

7. 声明两个变量，用来跟踪样本数与正确分类的样本数。

   ```
   int numberSamplesCorrect = 0; int numberSamples = 0;
   ```

8. 声明另外一个变量，用来跟踪分类耗费的时间。

   ```
   long evaluateStartTime =
     TimingUtils.getNanoCPUTimeOfCurrentThread();
   ```

9. 接下来，不断执行循环，直到数据流没有更多实例，并且分类的样本数达到总实例数。在循环中，先得到数据流每个实例的数据，而后查看分类器对实例所进行的分类是否正确。如果是，则把 `numberSamplesCorrect` 变量加 1。只有把测试功能打开（通过方法的第二个参数），才进行这种检查。

10. 然后，增加样本数，使用下一个训练实例训练学习器，进入下一次循环。

    ```
    while (stream.hasMoreInstances() && numberSamples < numInstances)
     {
      Instance trainInst = stream.nextInstance().getData();
      if (isTesting)
       {
        if (learner.correctlyClassifies(trainInst)){
         numberSamplesCorrect++;
        }
       }
      numberSamples++; learner.trainOnInstance(trainInst);
     }
    ```

11. 计算准确度。

    ```
    double accuracy = 100.0 * (double) numberSamplesCorrect/
      (double) numberSamples;
    ```

12. 并且计算分类耗费的时间。

```
double time = TimingUtils.nanoTimeToSeconds(TimingUtils.
    getNanoCPUTimeOfCurrentThread()- evaluateStartTime);
```

13. 最后,把这些评估指标显示出来,关闭方法。

```
System.out.println(numberSamples + " instances processed with "
    + accuracy + "% accuracy in "+time+" seconds."); }
```

14. 为了执行这个方法,应该为 main() 方法提供如下参数。最后关闭类。

```
public static void main(String[] args) throws IOException { MOA
    exp = new MOA(); exp.run(1000000, true); } }
```

示例的完整代码整理如下:

```
import moa.classifiers.trees.HoeffdingTree;
import moa.classifiers.Classifier;
import moa.core.TimingUtils;
import moa.streams.generators.RandomRBFGenerator;
import com.yahoo.labs.samoa.instances.Instance;
import java.io.IOException;

public class MOA {
        public void run(int numInstances, boolean isTesting){
                Classifier learner = new HoeffdingTree();
                RandomRBFGenerator stream = new RandomRBFGenerator();
                stream.prepareForUse();

                learner.setModelContext(stream.getHeader());
                learner.prepareForUse();

                int numberSamplesCorrect = 0;
                int numberSamples = 0;
                long evaluateStartTime = TimingUtils.
                  getNanoCPUTimeOfCurrentThread();
                while (stream.hasMoreInstances() && numberSamples <
                  numIn-stances) {
                        Instance trainInst =
                            stream.nextInstance().getData();
                        if (isTesting) {
                                if
                        (learner.correctlyClassifies(trainInst)){
                                numberSamplesCorrect++;
```

```
                    }
                }
                numberSamples++;
                learner.trainOnInstance(trainInst);
            }
            double accuracy = 100.0 * (double)
              numberSamplesCorrect/ (double) numberSamples;
            double time = Tim-in
            gUtils.nanoTimeToSeconds(TimingUtils.
               getNanoCPUTimeOfCurrentThread()- evaluateStartTime);
            System.out.println(numberSamples + " instances
             processed with " + accuracy + "% accuracy in "+time+"
               seconds.");
    }

    public static void main(String[] args) throws IOException {
        MOA exp = new MOA();
        exp.run(1000000, true);
    }
}
```

执行上面代码，得到如下输出结果（不同机器下输出结果可能不同）：

1000000 instances processed with 91.0458% accuracy in 6.769871032 seconds.

5.5 使用 Mulan 对多标签数据点进行分类

前面，我们学习了多类分类，其目标是把一个数据实例归入到某个类中。多标签数据实例是指那些拥有多个类或标签的数据实例。之前我们用过的机器学习工具不能处理这些拥有多个目标类的数据点。

在对多标签数据点进行分类时，我们将使用一个名为 Mulan 的开源 Java 库。Mulan 实现了各种分类、排序、特征选择、模型评估算法。Mulan 不带有 GUI 界面，你只能通过命令行或 API 来使用它。本部分，我们把学习的重点放在使用两种不同的分类器对多标签数据集进行分类并对分类结果进行评估上。

准备工作

开始之前，先做如下准备。

1. 首先，下载 Mulan。在本部分，我们将使用 Mulan 1.5 版本，请读者自行下载。

2. 下载完成后，将压缩文件解压缩，你将看到如下数据文件夹。找到并进入 `dist` 文件夹（见图 5-14）。

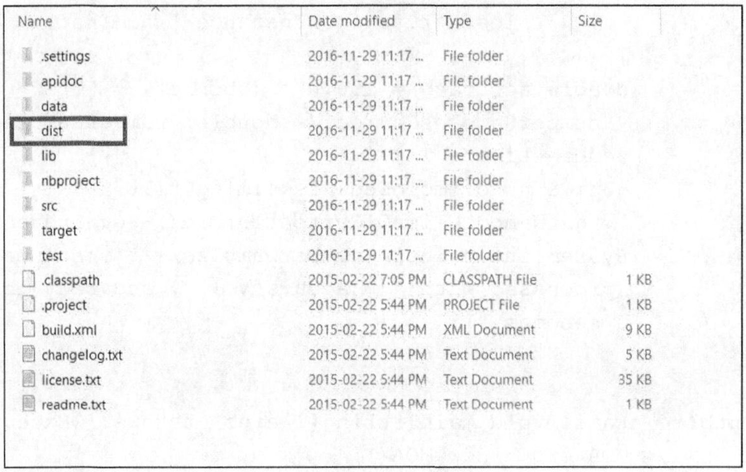

图 5-14

3. 如图 5-15 所示，在 `dist` 文件夹中，你会看到 3 个文件，其中有一个名为 `Mulan-1.5.0.zip` 的压缩文件，将其解压缩。

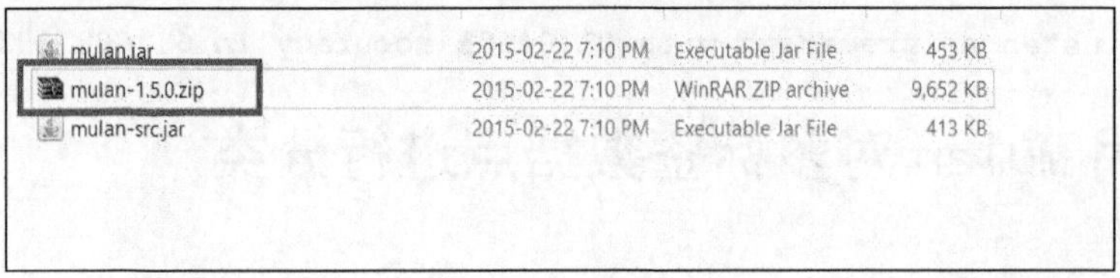

图 5-15

4. 解压之后，你会看到 3 个或 4 个 JAR 文件。在这 4 个 JAR 文件中，我们会用到其中 3 个，如图 5-16 所示。

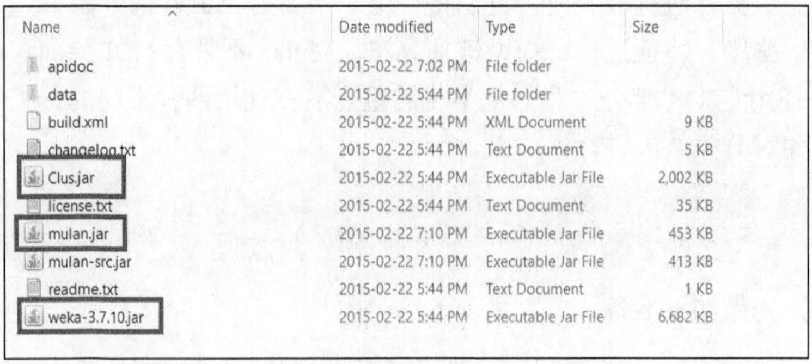

图 5-16

5. 如图 5-17 所示，把这 3 个 JAR 文件作为外部库添加到你的项目中。

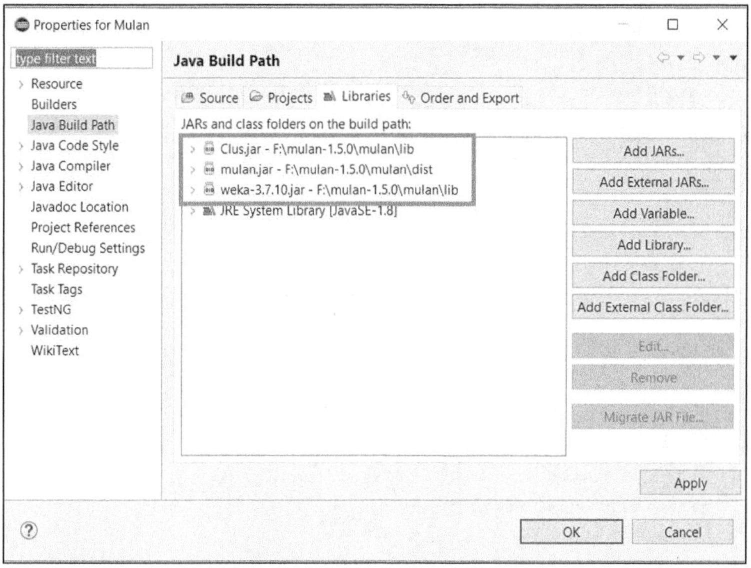

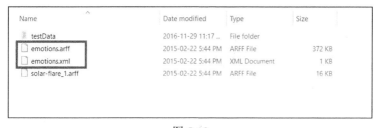

图 5-17

6. 在 Mulan 的 data 文件夹中包含着一个多标签数据集。如图 5-18 所示，在本部分，我们将使用 emotions.arff 与 emotions.xml 文件。

图 5-18

在正式开始操作之前，让我们先了解一个 Mulan 的数据集格式。Mulan 需要用到两个文件来指定一个多标签数据集。第一个文件是 ARFF 文件（有关 ARFF 文件的内容，请参考第 4 章）。标签应该是名义属性，带有两个值 0 与 1，前者表示标签不存在，后者表示标签存在。在下面这个例子中，数据集拥有 3 个数值特征，每个实例有 5 个类别或标签。在 @data 部分，前 3 个值表示实际特征值，5 个 0 或 1 分别表示类不存在或存在。

```
@relation MultiLabelExample

@attribute feature1 numeric
@attribute feature2 numeric
```

```
@attribute feature3 numeric
@attribute label1 {0, 1}
@attribute label2 {0, 1}
@attribute label3 {0, 1}
@attribute label4 {0, 1}
@attribute label5 {0, 1}

@data
2.3,5.6,1.4,0,1,1,0,0
```

另一个文件是 XML 文件,它用来指定标签及它们之间的层次关系。下面这个 XML 文件对应于上面的例子。

```
<labels xmlns="http://mulan.sourceforge.net/labels">
<label name="label1"></label>
   <label name="label2"></label>
   <label name="label3"></label>
   <label name="label4"></label>
   <label name="label5"></label>
</labels>
```

想了解更多细节,请访问 Mulan 官网。

操作步骤

1. 创建 Mulan 类与 main() 方法。我们会把所有代码写到 main() 方法之中。

    ```
    public class Mulan { public static void main(String[] args){
    ```

2. 创建一个数据集,并把 emotions.arff 与 emotions.xml 文件读入到数据集中。

    ```
    MultiLabelInstances dataset = null;
      try {
        dataset = new MultiLabelInstances("path to
          emotions.arff",
          "path to emotions.xml");
    } catch (InvalidDataFormatException e) {
    }
    ```

3. 接下来,创建 RAkEL 分类器与 MLkNN 分类器。请注意,RAkEL 是一个元分类器,它可以拥有一个多标签学习器,并且通常与 LabelPowerset 算法配用。

LabelPowerset 是一个基于变换的算法,它可以接收一个单标签分类器(本例中为 J48)作为参数。MLkNN 是一个自适应分类器,它基于 k 最近邻算法。

```
RAkEL learner1 = new RAkEL(new LabelPowerset(new J48()));
    MLkNN
  learner2 = new MLkNN();
```

4. 创建一个评估器,用来评估分类性能。

```
Evaluator eval = new Evaluator();
```

5. 由于我们会对两个分类器进行评估,所以需要声明一个变量,用来保存多个评估结果。

```
MultipleEvaluation results;
```

6. 进行评估时,我们会使用 10 折交叉验证。因此,声明一个变量,表示要创建的折数。

```
int numFolds = 10;
```

7. 接着,对第一个学习器进行评估,显示评估结果。

```
results = eval.crossValidate(learner1, dataset, numFolds);
System.out.println(results);
```

8. 最后,评估第二个学习器,并显示评估结果。

```
results = eval.crossValidate(learner2, dataset, numFolds);
System.out.println(results);
```

9. 关闭方法与类。

```
}}
```

示例的完整代码整理如下:

```
import mulan.classifier.lazy.MLkNN;
import mulan.classifier.meta.RAkEL;
import mulan.classifier.transformation.LabelPowerset;
import mulan.data.InvalidDataFormatException;
import mulan.data.MultiLabelInstances;
import mulan.evaluation.Evaluator;
import mulan.evaluation.MultipleEvaluation;
```

```
import weka.classifiers.trees.J48;

public class Mulan {
        public static void main(String[] args){
                MultiLabelInstances dataset = null;
                try {
                    dataset = new MultiLabelInstances("path to emotions.
                      arff", "path to emotions.xml");
                } catch (InvalidDataFormatException e) {
                }
                RAkEL learner1 = new RAkEL(new LabelPowerset(new
                  J48()));
                MLkNN learner2 = new MLkNN();
                Evaluator eval = new Evaluator();
                MultipleEvaluation results;
                int numFolds = 10;
                results = eval.crossValidate(learner1, dataset, num-
                  Folds);
                System.out.println(results);
                results = eval.crossValidate(learner2, dataset, num-
                  Folds);
                System.out.println(results);
        }
}
```

运行上面代码，输出对两个学习器的性能评估结果。

Fold 1/10
Fold 2/10
Fold 3/10
Fold 4/10
Fold 5/10
Fold 6/10
Fold 7/10
Fold 8/10
Fold 9/10
Fold 10/10
Hamming Loss: 0.2153±0.0251
Subset Accuracy: 0.2562±0.0481
Example-Based Precision: 0.6325±0.0547
Example-Based Recall: 0.6307±0.0560
Example-Based F Measure: 0.5990±0.0510
Example-Based Accuracy: 0.5153±0.0484

```
Example-Based Specificity: 0.8607±0.0213
.....................................
Fold 1/10
Fold 2/10
Fold 3/10
Fold 4/10
Fold 5/10
Fold 6/10
Fold 7/10
Fold 8/10
Fold 9/10
Fold 10/10
Hamming Loss: 0.1951±0.0243
Subset Accuracy: 0.2831±0.0538
Example-Based Precision: 0.6883±0.0655
Example-Based Recall: 0.6050±0.0578
Example-Based F Measure: 0.6138±0.0527
Example-Based Accuracy: 0.5326±0.0515
Example-Based Specificity: 0.8994±0.0271
.....................................
```

第 6 章
从文本数据提取信息

本章涵盖如下内容：

- 使用 Java 检测标记（单词）；
- 使用 Java 检测句子；
- 使用 OpenNLP 检测标记（单词）与句子；
- 使用 Stanford CoreNLP 从标记中提取词根与词性，以及识别命名实体；
- 使用 Java 8 借助余弦相似性测度（Cosine Similarity measure）测量文本相似度；
- 使用 Mallet 从文本文档提取主题；
- 使用 Mallet 对文本文档分类；
- 使用 Weka 对文本文档分类。

6.1 简介

随着互联网的发展，积累了大量 Web 数据，并且这些数据大部分都是文本格式的，这使得数据科学家处理的大部分数据类型是文本。文本数据来源非常多样，你可以从文档、文章、博客、社交媒体更新、新闻专线等多方面获取。

有许多基于 Java 的工具可供数据科学家用来从文本数据提取信息。此外，还有许多工具可以帮助我们做各种数据科学任务。本章，我们把讨论的范围限定在几个数据科学任务上，比如提取细小的文本特征（句子、单词）、使用机器学习进行文档分类、主题提取与建模、从文档中提取关键字以及命名实体识别。

6.2 使用 Java 检测标记（单词）

对数据科学家来说，一项最常见的任务是从文本数据中检测标记。这项任务叫作标记化（tokenization）。尽管"标记"可以是单词、符号、短语或其他有意义的文本单元（text unit），但在本章中，我们会把单词视作标记，因为单词本身就是可供处理的文本单元。然而，不同人对单词标记的理解不同，有些人只需要单词，有些人在检测时倾向于把符号忽略，而有些人想保留单词中的标点符号以便获取更多上下文。基于多样化的需求，本部分我们将对相同字符串应用 3 种不同的技术来产生 3 种不同的结果。这些技术包括字符串分隔、分词（BreakIterator）、正则表达式。至于选用哪种技术，需要你自己做决定。

请注意，这里我们只选择了 3 种方法，此外还有其他许多方法可用，这些方法留待你去探索吧。

准备工作

1. 了解 Pattern 类所支持的正则表达式模式的文档。
2. 查看示例，你可以大致了解 BreakIterator 的用法。

操作步骤

1. 首先，创建 useTokenizer 方法，它使用 Java 的 `StringTokenzier` 类来检测标记。这个方法接收句子输入，使用 `StringTokenzier` 类对句子进行切分。

   ```
   public void useTokenizer(String input){
   ```

2. 调用 `StringTokenizer` 构造函数，并把输入的句子作为参数传递给它。

   ```
   StringTokenizer tokenizer = new StringTokenizer(input);
   ```

3. 创建一个字符串对象来保存标记。

   ```
   String word ="";
   ```

4. 遍历 tokenizer 得到每个单词，并把它们在控制台上打印出来。

   ```
   while(tokenizer.hasMoreTokens()){
     word = tokenizer.nextToken();
   ```

```
System.out.println(word);
}
```

5. 关闭方法。

   ```
   }
   ```

   对于一个句子，比如 "Let's get this vis-a-vis, he said, "theseboys' marks are really that well? " "" "", 这个方法的输出如下：

   ```
   "Let's
   get
   this
   vis-a-vis",
   he
   said,
   "these
   boys'
   marks
   are
   really
   that
   well?"
   ```

6. 然后，我们创建 useBreakIterator 方法，它使用 Java 的 BreakIterator 类遍历文本中的每个单词。你会看到这个方法的代码比前面创建的第一个方法更加复杂一些。

   这个方法接收句子输入作为它的参数。

   ```
   public void useBreakIterator(String input){
   ```

7. 然后，使用 BreakIterator 类来创建 tokenizer 对象。

   ```
   BreakIterator tokenizer = BreakIterator.getWordInstance();
   ```

8. 向输入的句子应用 tokenizer。

   ```
   tokenizer.setText(input);
   ```

9. 获取 tokenizer 的起始索引。

```
          int start = tokenizer.first();
```

10. 使用 for 循环获取每个标记，从输入句子提取子字符串后，把它们在控制台打印出来。

```
    for (int end = tokenizer.next();
         end != BreakIterator.DONE;
         start = end, end = tokenizer.next()) {
        System.out.println(input.substring(start,end));
    }
```

11. 关闭方法。

```
    }
```

对于一个句子，比如 "Let's get this vis-a-vis, he said, "theseboys' marks are really that well? """"，这个方法的输出如下：

```
"
Let's
get
this
vis-a-vis
"
,
he
said
,
"
these
boys
'
marks
are
really
that
well
?
"
```

12. 最后，创建 useRegEx 方法，它使用正则表达式对输入文本进行切分。

```
public void useRegEx(String input){
```

13. 使用正则表达式的模式来捕获标点符号、单个或多个带有连字符的单词、引号、撇号等。如果你需要一些特定的模式，只要像下面这样使用你自己的正则表达式即可。

    ```
    Pattern pattern = Pattern.compile("\\w[\\w-]+('\\w*)?");
    ```

14. 在 pattern 上应用 matcher。

    ```
    Matcher matcher = pattern.matcher(input);
    ```

15. 使用 matcher 从输入文本提取所有单词。

    ```
    while ( matcher.find() ) {
      System.out.println(input.substring(matcher.start(),
        matcher.end()));
    }
    ```

16. 关闭方法。

    ```
    }
    ```

对于一个句子，比如 "Let's get this vis-a-vis, he said, "these boys' marks are really that well? " " " " "，这个方法的输出如下：

```
Let's
get
this
vis-a-vis
he
said
these
boys'
marks
are
really
that
well
```

示例的完整代码整理如下：

```java
import java.text.BreakIterator;
import java.util.StringTokenizer;
import java.util.regex.Matcher;
import java.util.regex.Pattern;

public class WordDetection {
    public static void main(String[] args){
        String input = ""Let's get this vis-a-vis", he said, "these boys'
          marks are really that well?"";
        WordDetection wordDetection = new WordDetection();
        wordDetection.useTokenizer(input);
        wordDetection.useBreakIterator(input);
        wordDetection.useRegEx(input);
    }
    public void useTokenizer(String input){
        System.out.println("Tokenizer");
        StringTokenizer tokenizer = new StringTokenizer(input);
        String word ="";
        while(tokenizer.hasMoreTokens()){
            word = tokenizer.nextToken();
            System.out.println(word);
        }
    }
    public void useBreakIterator(String input){
        System.out.println("Break Iterator");
        BreakIterator tokenizer = BreakIterator.getWordInstance();
        tokenizer.setText(input);
        int start = tokenizer.first();
        for (int end = tokenizer.next();
            end != BreakIterator.DONE;
            start = end, end = tokenizer.next()) {
            System.out.println(input.substring(start,end));
        }
    }
    public void useRegEx(String input){
        System.out.println("Regular Expression");
        Pattern pattern = Pattern.compile("\\w[\\w-]+('\\w*)?");
        Matcher matcher = pattern.matcher(input);
```

```
        while ( matcher.find() ) {
            System.out.println(input.substring(matcher.start(),
                matcher.end()));
        }
    }
}
```

6.3 使用 Java 检测句子

本部分，我们将学习检测句子的方法，在把它们检测出来之后，可以针对它们进行进一步分析。对数据科学家而言，句子是非常重要的文字单位，可以用来进行不同的练习，比如分类。为了从文本中检测出句子，我们将使用 Java 的 `BreakIterator` 类。

准备工作

查看示例，大致了解 BreakIterator 的用法。

操作步骤

为了测试本部分代码，我们将使用两个稍微特殊的句子，它们会使许多基于正则表达式的方法产生混淆。这两个句子是："My name is Rushdi Shams.You can use Dr. before my name as I have a PhD. but I am a bit shy to use it." 有趣的是，我们使用 Java 的 `BreakIterator` 类可以高效地处理它们。

创建名为 useSentenceIterator 的方法，其参数是测试字符串。

```
        public void useSentenceIterator(String source){
```

1. 创建 `BreakIterator` 的对象 `iterator`。

    ```
    BreakIterator iterator =
      BreakIterator.getSentenceInstance(Locale.US);
    ```

2. 对测试字符串应用这个 `iterator`。

    ```
    iterator.setText(source);
    ```

3. 获取测试字符串的起始索引，并将其存储在一个整型变量中。
    ```
    int start = iterator.first();
    ```

4. 最后，遍历 iterator 中的所有句子，并把它们打印出来。为了遍历 iterator 中的所有句子，我们需要另一个名为 end 的变量来指向句子的结束索引。

```
for (int end = iterator.next(); end != BreakIterator.DONE;
  start = end, end = iterator.next()) {
  System.out.println(source.substring(start,end));
}
```

代码输出结果如下：

```
My name is Rushdi Shams.
You can use Dr. before my name as I have a Ph.D. but I am a bit shy to use it.
```

示例的完整代码整理如下：

```
import java.text.BreakIterator;
import java.util.Locale;
public class SentenceDetection {
  public void useSentenceIterator(String source){
    BreakIterator iterator =
      BreakIterator.getSentenceInstance(Locale.US);
    iterator.setText(source);
    int start = iterator.first();
    for (int end = iterator.next();
      end != BreakIterator.DONE;
      start = end, end = iterator.next()) {
      System.out.println(source.substring(start,end));
    }
  }
  public static void main(String[] args){
    SentenceDetection detection = new SentenceDetection();
    String test = "My name is Rushdi Shams. You can use Dr. before my name as I have a Ph.D. but I am a bit shy to use it.";
    detection.useSentenceIterator(test);
  }
}
```

6.4 使用 OpenNLP 检测标记（单词）与句子

本章前面两节讲解了使用传统 Java 类与方法检测标记（单词）与句子。本节，我们将使用一个名为 Apache OpenNLP 的开源库把检测标记与句子的两项任务组合起来。其实，这两项任务使用传统的方法也能得到很好的解决，这里之所以要使用 OpenNLP，目的是向各位数据科学家介绍这个工具，它用起来非常方便，并且在针对标准与传统的语料库执行一些信息提取任务时有着更高的准确度。在使用这个库进行词语切分、断句、词性标注、命名实体识别、分块、语法分析与指代消解时，你可以根据自己的文章或文档的语料库训练分类器。

准备工作

1. 写作本书之时，OpenNLP 最新版本为 1.6.0，推荐你使用这个版本（见图 6-1）。下载 OpenNLP 1.6.0。前往这个页面，下载二进制 zip 文件。

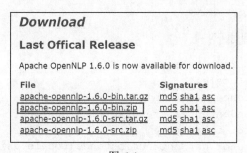

图 6-1

2. 下载完成后，解压缩文件。如图 6-2 所示，在解压之后的目录中，你将看到一个名称为 lib 的文件夹。

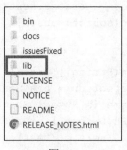

图 6-2

3. 如图 6-3 所示，在 lib 文件夹中，你会看到如下两个 Jar 文件。

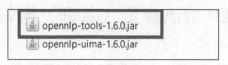

图 6-3

如图 6-4 所示，选择 `opennlp-tools-1.6.0.jar` 文件，将其作为外部库添加到你为本节创建的 Eclipse 项目中。

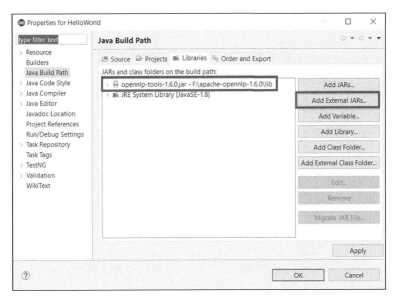

图 6-4

在本节中，我们将使用 OpenNLP 所提供的预建模型来进行标记与句子检测任务。因此，需要你先把模型下载并保存到本地硬盘上。请记住这些模型的保存路径，方便你把它们添加到代码中。

如图 6-5 所示，下载英文分词器与句子检测器模型。把它们保存在 C:/目录下。接下来，我们要准备写一些代码。

de	POS Tagger	Perceptron model trained on tiger corpus.	de-pos-perceptron.bin
en	Tokenizer	Trained on opennlp training data.	en-token.bin
en	Sentence Detector	Trained on opennlp training data.	en-sent.bin
en	POS Tagger	Maxent model with tag dictionary.	en-pos-maxent.bin

图 6-5

操作步骤

1. 本部分，我们将创建 useOpenNlp 方法，它使用 OpenNLP 的分词器（`tokenizer`）与句子检测器模型来进行分词，把原文本切分为多个句子。useOpenNlp 方法带有如下 3 个参数。

- 包含源文本的字符串。
- 模型路径。
- 这个字符串用来指明你想如何切分源文本，是把它们切分成一个个的词，还是把它们切分成句子。若是前者，请把参数指定为 word，若是后者，请把参数指定为 sentence。

```
public void useOpenNlp(String sourceText, String modelPath,
    String choice) throws IOException{
```

2. 首先，以输入流方式读取模型。

```
InputStream modelIn = null;
modelIn = new FileInputStream(modelPath);
```

3. 然后，针对 choice 为 sentence 的情形，创建一个 if 语句块，其中包含的代码用来从源文本检测句子片段。

```
if(choice.equalsIgnoreCase("sentence")){
```

4. 根据预建模型创建一个句子模型，然后关闭，用来保存预建模型的变量。

```
SentenceModel model = new SentenceModel(modelIn);
modelIn.close();
```

5. 借助刚刚创建的模型，创建一个句子检测器。

```
SentenceDetectorME sentenceDetector = new
    SentenceDetectorME(model)
```

6. 使用句子检测器从源文本检测句子。并把得到的结果保存在一个字符串数组中。

```
String sentences[] = sentenceDetector.sentDetect(sourceText);
```

7. 接着，把句子打印在控制台中，关闭 if 语句块。

```
System.out.println("Sentences: ");
    for(String sentence:sentences){
        System.out.println(sentence);
    }
}
```

8. 然后，创建一个 else if 语句块，其中包含着用来对源文本进行词语拆分的代码。

    ```
    else if(choice.equalsIgnoreCase("word")){
    ```

9. 使用预建模型创建一个 `tokenizer` 模型，关闭预建模型。

    ```
    TokenizerModel model = new TokenizerModel(modelIn);
    modelIn.close();
    ```

10. 使用刚创建的模型，创建一个 `tokenizer`。

    ```
    Tokenizer tokenizer = new TokenizerME(model);
    ```

11. 使用该 `tokenizer`，从源文本提取标记（单词），并把它们存入一个字符串数组中。

    ```
    String tokens[] = tokenizer.tokenize(sourceText);
    ```

12. 最后，在控制台中打印出标记，并关闭 else if 语句块。

    ```
    System.out.println("Words: ");
      for(String token:tokens){
          System.out.println(token);
      }
    }
    ```

13. 我们还需要编写一个 else 语句块，用来处理用户可能做出的无效选择。

    ```
    else{
      System.out.println("Error in choice");
      modelIn.close();
      return;
    }
    ```

14. 关闭方法。

    ```
    }
    ```

示例的完整代码整理如下：

```
import java.io.FileInputStream;
import java.io.IOException;
import java.io.InputStream;

import opennlp.tools.sentdetect.SentenceDetectorME;
import opennlp.tools.sentdetect.SentenceModel;
import opennlp.tools.tokenize.Tokenizer;
import opennlp.tools.tokenize.TokenizerME;
```

```java
import opennlp.tools.tokenize.TokenizerModel;

public class OpenNlpSenToken {
    public static void main(String[] args){
        OpenNlpSenToken openNlp = new OpenNlpSenToken();
        try {
            openNlp.useOpenNlp("My name is Rushdi Shams. "
                + "You can use Dr. before my name as I have a Ph.D. "
                + "but I am a bit shy to use it.", "C:/en-sent.bin",
                    "sentence");
            openNlp.useOpenNlp("\"Let's get this vis-a-vis\", he said,
                "these boys' marks are really that well?\"", "C:/entoken.
                    bin", "word");
        } catch (IOException e) {
        }
    }
    public void useOpenNlp(String sourceText, String modelPath, String
        choice) throws IOException{
        InputStream modelIn = null;
        modelIn = new FileInputStream(modelPath);

        if(choice.equalsIgnoreCase("sentence")){
            SentenceModel model = new SentenceModel(modelIn);
            modelIn.close();
            SentenceDetectorME sentenceDetector = new
            SentenceDetectorME(model);
            String sentences[] = sentenceDetector.sentDetect(sourceText);
            System.out.println("Sentences: ");
            for(String sentence:sentences){
                System.out.println(sentence);
            }
        }
        else if(choice.equalsIgnoreCase("word")){
            TokenizerModel model = new TokenizerModel(modelIn);
            modelIn.close();
            Tokenizer tokenizer = new TokenizerME(model);
            String tokens[] = tokenizer.tokenize(sourceText);
            System.out.println("Words: ");
            for(String token:tokens){
                System.out.println(token);
            }
        }
        else{
```

```
            System.out.println("Error in choice");
            modelIn.close();
            return;
        }
    }
}
```

此时，你可以把上面代码的输出与前两节的输出进行比较，它们所使用的源文本都是一样的。

  关于OpenNLP库的其他用法，强烈建议你参阅其官网。

6.5 使用Stanford CoreNLP从标记中提取词根、词性，以及识别命名实体

前面我们学习了从给定文本提取标记或单词的方法，接下来我们将学习如何从这些标记获取不同信息，比如它们的词根、词性，以及指定的标记是否是一个命名实体。

词形还原处理是将一个单词被改变的各种形式归为一类，可以把它们作为一个单独的文本单元进行分析。这与词干提取处理类似，不同之处在于归类过程中词干提取不考虑上下文。因此，就文本数据分析来说，相比于词干提取，词形还原更有用，但是执行这项任务通常需要使用更多的计算能力。

文章或文档中标记的词性标签在许多机器学习模型中被广泛用作特征，这对数据科学家非常有用。

另一方面，命名实体对于分析新闻文章数据非常重要，对有关商业公司的研究有着巨大影响。

本部分，我们将学习使用Standford CoreNLP 3.7.0（写作本章时的最新版本）从文本提取这些信息的方法。

准备工作

1. 下载Stanford CoreNLP 3.7.0。

2. 下载完成后，把下载得到的文件进行解压缩，你会看到如图6-6所示的目录结构。

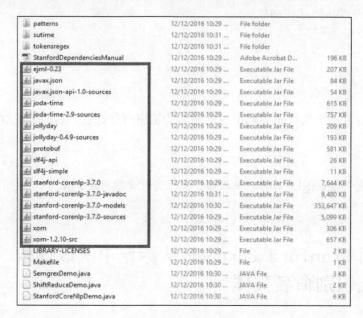

图6-6

3. 把图6-6框线中的所有Jar文件作为外部Jar文件添加到你的现有项目中，如图6-7所示。至此，编写代码前的准备工作就做好了。

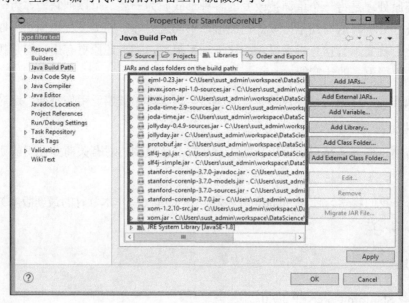

图6-7

操作步骤

1. 创建 Lemmatizer 类与 main() 方法，我们将把本节的所有代码全部写在 main() 方法中。

   ```
   public class Lemmatizer {
     public static void main(String[] args){
   ```

2. 接着，创建 Stanford CoreNLP 流水线。通过这个流水线，你可以把多个属性值提供给 CoreNLP 引擎。

   ```
   StanfordCoreNLP pipeline;
   ```

3. 创建一个 Properties 对象，向它添加几个属性。本例中，我们将使用词性标注与词形还原来进行标记。

   ```
   Properties props = new Properties();
   props.put("annotators", "tokenize, ssplit, pos, lemma, ner");
   ```

4. 接着，使用这些属性，创建一个 CoreNLP 对象。

   ```
   pipeline = new StanfordCoreNLP(props, false);
   ```

5. 创建字符串，后面我们将为它生成词根。

   ```
   String text = "Hamlet's mother, Queen Gertrude, says this
       famous line while watching The Mousetrap. "
         + "Gertrude is talking about the queen in the play. "
         + "She feels that the play-queen seems insincere because
           she repeats so dramatically that she'll never remarry
           due to her undying love of her husband.";
   ```

6. 接着，使用给定的文本创建一个 Annotation。

   ```
   nnotation document = pipeline.process(text);
   ```

7. 最后，为每个标记获取原始标记与标记词根。其实，你不必获取原始标记，但是为了观察单词形式与词根形式的不同，我们有意这样做，也很容易办到。使用名为 document 的 Annotation 变量（该变量已经在上一步创建），对所有句子重复这个过程。

```
        for(CoreMap sentence: document.get(SentencesAnnotation.class))
        {
            for(CoreLabel token: sentence.get(TokensAnnotation.class))
            {
                String word = token.get(TextAnnotation.class);
                String lemma = token.get(LemmaAnnotation.class);
                String pos = token.get(PartOfSpeechAnnotation.class);
                String ne = token.get(NamedEntityTagAnnotation.class);
                System.out.println(word + "-->" + lemma + "-->" + pos
                + "-->" + ne);
            }
        }
```

8. 关闭方法与类。

```
    }
}
```

上面代码的部分输出如下:

```
...
Queen-->Queen-->NNP-->PERSON
Gertrude-->Gertrude-->NNP-->PERSON
,-->,-->,-->O
says-->say-->VBZ-->O
this-->this-->DT-->O
famous-->famous-->JJ-->O
line-->line-->NN-->O
while-->while-->IN-->O
watching-->watch-->VBG-->O
The-->the-->DT-->O
Mousetrap-->mousetrap-->NN-->O
.-->.-->.-->O
Gertrude-->Gertrude-->NNP-->PERSON
is-->be-->VBZ-->O
talking-->talk-->VBG-->O
...
```

示例的完整代码整理如下:

```
import edu.stanford.nlp.ling.CoreAnnotations.LemmaAnnotation;
import edu.stanford.nlp.ling.CoreAnnotations.NamedEntityTagAnnotation;
import edu.stanford.nlp.ling.CoreAnnotations.PartOfSpeechAnnotation;
```

```
import edu.stanford.nlp.ling.CoreAnnotations.SentencesAnnotation;
import edu.stanford.nlp.ling.CoreAnnotations.TextAnnotation;
import edu.stanford.nlp.ling.CoreAnnotations.TokensAnnotation;
import edu.stanford.nlp.ling.CoreLabel;
import edu.stanford.nlp.pipeline.Annotation;
import edu.stanford.nlp.pipeline.StanfordCoreNLP;
import edu.stanford.nlp.util.CoreMap;
import java.util.Properties;

public class Lemmatizer {
    public static void main(String[] args){
      StanfordCoreNLP pipeline;
        Properties props = new Properties();
        props.put("annotators", "tokenize, ssplit, pos, lemma, ner");
        pipeline = new StanfordCoreNLP(props, false);
        String text = "Hamlet's mother, Queen Gertrude, says this
          famous line while watching The Mousetrap. "
            + "Gertrude is talking about the queen in the play. "
            + "She feels that the play-queen seems insincere because
              she repeats so dramatically that she'll never remarry
                due to her undying love of her husband.";
        Annotation document = pipeline.process(text);

        for(CoreMap sentence: document.get(SentencesAnnotation.class))
          {
            for(CoreLabel token: sentence.get(TokensAnnotation.class))
             {
                String word = token.get(TextAnnotation.class);
                String lemma = token.get(LemmaAnnotation.class);
                String pos = token.get(PartOfSpeechAnnotation.class);
                String ne = token.get(NamedEntityTagAnnotation.class);
                System.out.println(word + "-->" + lemma + "-->" + pos
                + "-->" + ne);
             }
          }
     }
}
```

6.6 使用 Java 8 借助余弦相似性测度测量文本相似度

在进行分类、聚类、检测异常值，以及其他许多情形下，数据科学家常常需要对两个数据点之间的距离或相似度进行测量。当处理的数据点是文本时，我们就无法使用传统的距离或相似度测量方式。有很多标准、经典的相似性度量方法（也包括很多新出现的）可以用来比较两个或多个文本数据点。本部分，我们将使用余弦相似度（Cosine Similarity）来计算两个句子的距离。在信息提取中，余弦相似度是一个事实上的标准，应用广泛。本节中，我们将使用它来测量两个句子（字符串格式）的相似度。

准备工作

若想全面了解有关余弦相似度的内容，你可以参阅维基百科的介绍，下面让我们一起看一下如何对两个句子应用这个算法。

1. 首先，从两个字符串提取单词。
2. 为各个字符串中的每个单词，计算出现频率。这里所说的"出现频率"是指单词在每个句子中出现的次数。设 A 为单词向量，表示它们在第一个字符串出现的频率，B 也是单词向量，表示它们在第二个字符串中出现的频率。
3. 通过删除重复项，找出每个字符串的特有单词。
4. 找出两个字符串都有的单词。
5. 余弦相似度公式的分子是向量 A 与 B 的点积。
6. 公式的分母是向量 A 与 B 大小的算术积。

> 请注意，两个句子的余弦相似度分数介于 –1（表示正好相反）到 1（表示完全一样）之间，0 分表示无关。

操作步骤

1. 创建 `calculateCosine` 方法，它带有两个字符串参数，我们将为这两个字符串计算余弦相似度。

    ```
    public double calculateCosine(String s1, String s2){
    ```

2. 我们将使用正则表达式与 Java 8 的并行功能对给定的字符串进行切分。最终得到单词的两个字符串流。

    ```
    Stream<String> stream1 =
      Stream.of(s1.toLowerCase().split("\\W+")).parallel();
    Stream<String> stream2 =
    ```

```
Stream.of(s2.toLowerCase().split("\\W+")).parallel();
```

 进行切分处理时，你可以使用上一章第一节中介绍的任何一种方法，但是这一步所使用的方法也很方便、简洁，它充分利用了正则表达式与 Java 8 的功能。

3. 接着，使用 Java 8 获取每个字符串中每个单词出现的频率。最后，你会得到两个 Map。

```
Map<String, Long> wordFreq1 = stream1
  .collect(Collectors.groupingBy
  (String::toString,Collectors.counting()));
Map<String, Long> wordFreq2 = stream2
  .collect(Collectors.groupingBy
  (String::toString,Collectors.counting()));
```

4. 对于每个句子的单词列表，删除其中的重复项，仅保留唯一的单词。为此，我们需要创建两个集合，它们含有上一步中创建的 Map 对象。

```
Set<String> wordSet1 = wordFreq1.keySet();
Set<String> wordSet2 = wordFreq2.keySet();
```

5. 为了把步骤 3 中两个 Map 的点积（在余弦相似度计算中用作分子）计算出来，我们需要为两个字符串共有的单词创建列表。

```
Set<String> intersection = new HashSet<String>(wordSet1);
intersection.retainAll(wordSet2);
```

6. 接着，计算公式的分子，它是两个 Map 的点积。

```
double numerator = 0;
  for (String common: intersection){
  numerator += wordFreq1.get(common) * wordFreq2.get(common);
}
```

7. 接下来，我们将计算公式的分母，它是两个 Map 大小的算术积。

   首先，创建两个变量，用来保存向量大小的值（存在于 Map 数据结构中）。

```
double param1 = 0, param2 = 0;
```

8. 然后，计算第一个向量的大小。

```
for(String w1: wordSet1){
  param1 += Math.pow(wordFreq1.get(w1), 2);
}
param1 = Math.sqrt(param1);
```

9. 接着,计算第二个向量的大小。

```
for(String w2: wordSet2){
  param2 += Math.pow(wordFreq2.get(w2), 2);
}
param2 = Math.sqrt(param2);
```

10. 现在,我们已经得到了两个向量的大小,把它们乘起来,即可计算出分母。

```
double denominator = param1 * param2;
```

11. 最后,计算出两个字符串的余弦相似度,并返回给调用者,然后关闭方法。

```
double cosineSimilarity = numerator/denominator;
return cosineSimilarity;
}
```

12. 示例的完整代码整理如下:

```
import java.util.HashSet;
import java.util.Map;
import java.util.Set;
import java.util.stream.Collectors;
import java.util.stream.Stream;

public class CosineSimilarity {
  public double calculateCosine(String s1, String s2){
    //使用Java 8 并行分词
    Stream<String> stream1 =
      Stream.of(s1.toLowerCase().split("\\W+")).parallel();
    Stream<String> stream2 =
      Stream.of(s2.toLowerCase().split("\\W+")).parallel();
    //两个字符串的词频 Map
    Map<String, Long> wordFreq1 = stream1
      .collect(Collectors.groupingBy
        (String::toString,Collectors.counting()));
    Map<String, Long> wordFreq2 = stream2
      .collect(Collectors.groupingBy
```

```
            (String::toString,Collectors.counting()));
        //每个字符串的特有单词
        Set<String> wordSet1 = wordFreq1.keySet();
        Set<String> wordSet2 = wordFreq2.keySet();
        //两个字符串的共用单词
        Set<String> intersection = new HashSet<String>(wordSet1);
        intersection.retainAll(wordSet2);
        //余弦公式的分子 s1.s2
        double numerator = 0;
        for (String common: intersection){
            numerator += wordFreq1.get(common) * wordFreq2.get(common);
        }
        //余弦公式分母的两个参数
        double param1 = 0, param2 = 0;
        //sqrt(s1词频的平方和)
        for(String w1: wordSet1){
            param1 += Math.pow(wordFreq1.get(w1), 2);
        }
        param1 = Math.sqrt(param1);
        // sqrt(s2词频的平方和)
        for(String w2: wordSet2){
            param2 += Math.pow(wordFreq2.get(w2), 2);
        }
        param2 = Math.sqrt(param2);
        //余弦公式分母 sqrt(sum(s1^2))X
            sqrt(sum(s2^2))
        double denominator = param1 * param2;
        //计算余弦相似度
        double cosineSimilarity = numerator/denominator;
        return cosineSimilarity;
    }//关闭用来计算两个字符串余弦相似度的方法。
    public static void main(String[] args){
        CosineSimilarity cos = new CosineSimilarity();
        System.out.println(cos.calculateCosine("To be, or not to be: that
            is the question.", "Frailty, thy name is woman!"));
        System.out.println(cos.calculateCosine("The lady doth protest too
            much, methinks.", "Frailty, thy name is woman!"));
    }
}
```

运行上面代码,得到如下输出:

0.11952286093343936

```
0.0
```

输出结果表明"To be, or not to be: that is the question."与"Frailty, thy name is woman!"两个句子之间的余弦相似度大约为 `0.11`;"The lady doth protest too much, methinks."与"Frailty, thy name is woman!"之间的余弦相似度为 `0.0`。

本节中,我们并没有把字符串中的停用词(stop words)删除。从两个文本单元中移除停用词更有助于我们得到无偏输出结果。

6.7 使用 Mallet 从文本文档提取主题

随着文本文档数量的不断增长,数据科学家面临的一项重要任务是从大量文章获得概述,包括提取摘要、总结、主题列表等。之所以这样做,并不是为了节省阅读这些文章的时间,而是为了进行聚类、分类、语义相关测量、情绪分析等。

在机器学习与自然语言处理领域,主题建模是指使用统计模型从文字文章提取摘要主题或关键字。本部分,我们将使用一个基于 Java 的机器学习与自然语言处理库——Mallet 来做这项任务,Mallet 是 Machine Learning for Language Toolkit 的缩写,它广泛应用于学术研究与工业领域,主要用来做如下任务:

- 文档分类;
- 聚类;
- 主题建模;
- 信息提取。

在上面这些领域中,本书只讲解主题建模与文档分类相关的内容。本节我们将学习使用 Mallet 提取主题的方法,而下一节我们主要学习使用 Mallet 借助监督机器学习的方法对文本文档进行分类。

请注意,在本节与接下来的章节中,我们将只通过命令行来使用 Mallet,而不会涉及编写代码的内容。这是因为借助命令行我们能更方便地使用 Mallet。如果你想知道如何通过 Java API 来使用 Mallet,建议你阅读 Mallet 丰富的 API 文档。

准备工作

1. 首先，安装 Mallet。本节中，我们只讲解如何在 Windows 操作系统中安装 Mallet。下载 Mallet，在本书写作时，Mallet 最新版本是 2.0.8，建议下载 Mallet 2.0.8（最好下载 zip 文件，如图 6-8 所示）。

图 6-8

2. 下载完成后，解压缩到 C:/ 目录中。如图 6-9 所示，在你的 C:/ 目录之下将出现一个名为 C:\mallet-2.0.8RC2 的目录。

图 6-9

3. 在 C:\mallet-2.0.8RC2 目录之下，包含如图 6-10 所示的文件与文件夹。其中可实际运行的文件位于 bin 文件夹中，一些示例数据集包含在 sample-data 文件夹中。

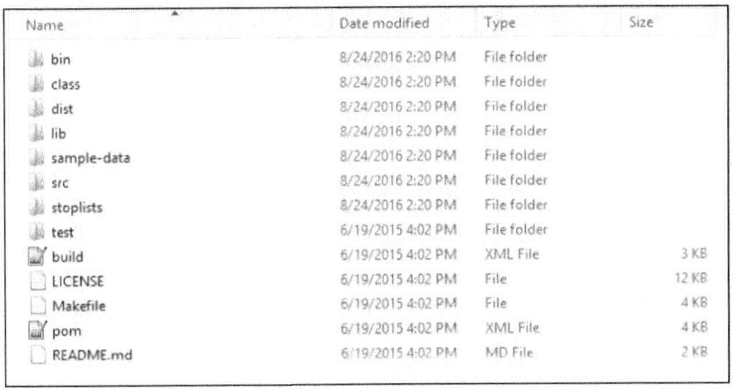

图 6-10

4. 在你的 Windows PC 中，进入"控制面板\系统与安全\系统"，单击"高级系统设置"（如图 6-11 所示）。

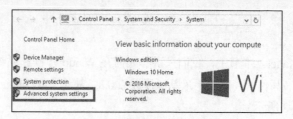

图 6-11

5. 在弹出的"系统属性"对话框中,单击"环境变量"按钮(如图 6-12 所示)。

图 6-12

6. 在打开的"环境变量"对话框中,在"系统变量"下单击"新建"(如图 6-13 所示)。

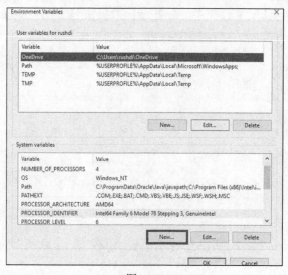

图 6-13

7. 在"新建系统变量"对话框中,在"变量名"中输入"MALLET_HOME",在"变量值"中,输入 C:\mallet-2.0.8RC2,单击"确定"按钮,关闭对话框。

8. 为了查看 Mallet 是否正确安装,打开命令行窗口,进入 Mallet 的 bin 目录下(见图 6-14),敲入 mallet 命令,在屏幕上将显示出 Mallet 2.0 中所有可用的命令。

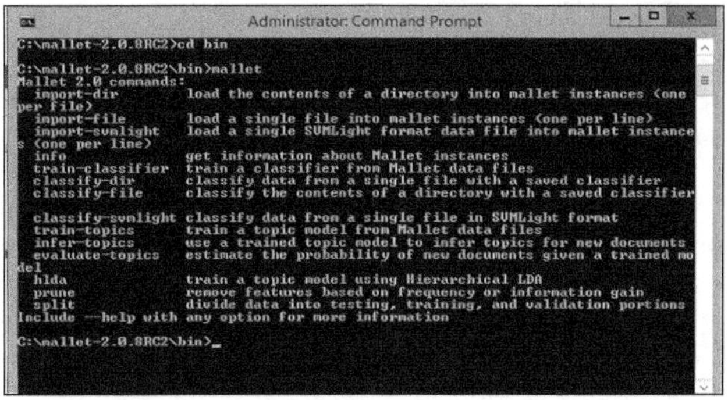

图 6-14

现在,你可以正常使用 Mallet 了。如图 6-15 所示,使用 Mallet 时,如果你不确定有哪些命令与选项可用,你可以使用帮助命令把 Mallet 2.0 中的所有可用命令与选项列出来。

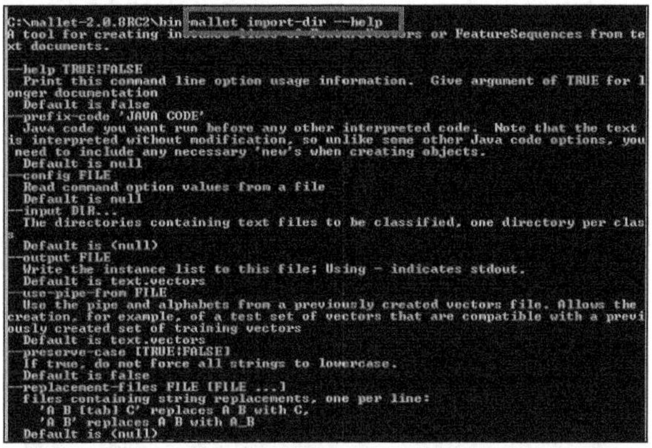

图 6-15

操作步骤

1. 在 Mallet 目录(位于 C:/之下)中包含一个名为 sample-data 的子目录。该目录下包含一个名为 Web 的目录。在 Web 目录中包含一个名为 en 的目录(里面包含几个文本文件,它们是一些英文文章)与另一个名为 de 的目录(其中包含几个文本文件,它们是一些德文文章)。本节中,我们将把 en 目录用作数据集或语料库,从其中的文章中提取主题。如果你想从自己准备的文档中提取主题时，下面这些操作步骤同样也是适用的，只需把 en 目录替换成包含你的文档的目录即可。

为此，首先把文本文件转换成单个文件，这种文件是一种 Mallet 文件类型，是一种二进制形式，人类不可读。在命令行窗口中，进入 C:\mallet-2.0.8RC2/bin 目录下，输入如下命令：

```
mallet import-dir --input C:\mallet-2.0.8RC2\sample
  data\web\en --output c:/web.en.mallet --keep-sequence --
  remove-stopwords
```

执行上面命令，将在 C:/ 中生成一个名为 web.en.mallet 的 Mallet 文件，里面保存着数据文件的原始序列（与 en 目录中的一样），并且从它的标准英文字典中删除停用词。

如果你想在建模期间让模型兼顾文本的二元文法（bi-grams），请使用如下命令代替上面的命令：

```
mallet import-dir --input C:\mallet-2.0.8RC2\sample- data\web\en
--output c:/web.en.mallet --keep-sequencebigrams --remove-stopwords
```

2. 输入如下命令，针对 web.en.mallet 文件，执行 Mallet 主题建模例程，默认设置保持不变。

```
mallet train-topics --input c:/web.en.mallet
```

执行上面命令，将在命令行中产生如图 6-16 所示的信息。

图 6-16

让我们检查一下输出。Mallet 主题建模输出的第二行包含如下内容：

```
1       5       zinta role hindi actress film indian acting
   top-grossing naa awa
rd female filmfare earned films drama written areas evening
    protection april
```

如果你是一个北印度片影迷，那么你会立刻意识到主题是由普丽缇·泽塔主演的北印度电影。为了确认这一点，你可以查看 C:\mallet-2.0.8RC2\sampledata\web\en 目录中名为 zinta.txt 的文件。

输出中的"1"表示段落编号（从 0 开始编号），"5"是主题的狄利克雷参数（Dirichlet parameter，它被视作主题的权重）。由于我们未进行设置，输出中所有段落的编号都是默认的。

 在使用 Mallet 进行主题建模与提取的过程中存在不确定性因素，因此即使针对相同的数据集，程序每次运行产生的关键字列表看上去也是不同的。当你看到的输出结果与上面的输出结果不同时，不要认为发生了错误。

这一步的命令十分通用，它不使用 Mallet 的任何参数，用来在控制台中显示结果。

3. 接下来，我们将把主题模型应用到相同数据上，但是使用了更多选项，并且把主题输出到一个外部文件中，以便我们进一步使用它们。在命令行中，输入如下命令：

```
mallet train-topics --input c:/web.en.mallet --num-topics 20--
    num-top-words 20 --optimize-interval 10 --xml-topic-phrase-
        report C:/web.en.xml
```

在这个命令中，我们把 c:/web.en.mallet 文件用作输入，为数据最多生成 20 个主题，把 20 个主题打印出来，并把结果输出到 c:/web.en.xml 文件。使用 --optimize-interval 选项打开超参数优化，用来生成更好的主题模型，它通过对一些主题进行优先排序让模型可以更好地拟合数据。

执行上面命令之后，在 C:/ 目录下，你将看到一个名为 web.en.xml 的 XML 文件。打开这个文件，你会看到如图 6-17 所示的内容。

```
<?xml version='1.0' ?>
<topics>
  <topic id="0" alpha="2.032473408279035" totalTokens="66" titles="test, paper,
    <word weight="0.045454545454545456" count="3">test</word>
    <word weight="0.030303030303030304" count="2">paper</word>
    <word weight="0.030303030303030304" count="2">played</word>
    <word weight="0.015151515151515152" count="1">regular</word>
    <word weight="0.015151515151515152" count="1">markets</word>
    <word weight="0.015151515151515152" count="1">commercial</word>
    <word weight="0.015151515151515152" count="1">fiction</word>
    <word weight="0.015151515151515152" count="1">kya</word>
    <word weight="0.015151515151515152" count="1">female</word>
    <word weight="0.015151515151515152" count="1">dil</word>
    <word weight="0.015151515151515152" count="1">made</word>
    <word weight="0.015151515151515152" count="1">films</word>
    <word weight="0.015151515151515152" count="1">hindi</word>
    <word weight="0.015151515151515152" count="1">actress</word>
    <word weight="0.015151515151515152" count="1">caused</word>
    <word weight="0.015151515151515152" count="1">standards</word>
    <word weight="0.015151515151515152" count="1">editor</word>
    <word weight="0.015151515151515152" count="1">graduated</word>
    <word weight="0.015151515151515152" count="1">robert</word>
    <word weight="0.015151515151515152" count="1">telescope</word>
    <phrase weight="0.25" count="1">telescope number</phrase>
    <phrase weight="0.25" count="1">australian cricketers</phrase>
    <phrase weight="0.25" count="1">hindi films</phrase>
    <phrase weight="0.25" count="1">test cricket-a</phrase>
</topic>
```

图 6-17

4. Mallet 中还有其他一些选项可以用在主题建模中。alpha 参数是其中最重要的一个选项，它也称为主题分布的平滑参数。在命令行中，尝试输入如下命令。

```
mallet train-topics --input c:/web.en.mallet --num-topics 20--
    num-top-words 20 --optimize-interval 10 --alpha 2.5 --xml-
    topic-phrase-report C:/web.en.xml
```

根据经验，一般把 alpha 值设置为 50/T，其中 T 是所选主题的数量，它通过 num-topics [数字]选项进行设定。如果要生成 20 个主题，那么我们应该把 alpha 值设置为 50/20=2.5。

在为一个文档创建主题模型时，若没有设置 random-seed，则会采用随机值，每次生成的包含主题的 xml 文件都不同，或大或小程度不同。

5. Mallet 还可以按不同格式输出主题，帮助用户采用不同方式分析主题。在命令行中输入如下命令：

```
mallet train-topics --input c:/web.en.mallet --num-topics 20--
    num-top-words 20 --optimize-interval 10 --output-state
      C:\web.en.gz --output-topic-keys C:\web.en.keys.txt --
        output-doc-topics c:/web.en.composition.txt
```

执行上面命令后，将在 C:/ 下生成 3 个新文件。

- `C:\web.en.gz` 包含一个文件，其中有语料库中的每个单词及其所属主题。该文件的部分内容显示如图 6-18 所示。

```
#doc source pos typeindex type topic
#alpha : 0.00529288685745327 0.005280224930316231 0.010411314990479141 0.005292639
0.005278782954717919 3.8154450835853604E-56 8.363084051683433E-65 0.01080648155617
0.005286647403862656 1.0743671245908252E-50 1.8112326797320508E-59 1.0996963864009
#beta : 0.2613020959658213
0 C:\mallet-2.0.8RC2\sample-data\web\en\elizabeth_needham.txt 0 0 elizabeth 2
0 C:\mallet-2.0.8RC2\sample-data\web\en\elizabeth_needham.txt 1 1 needham 2
0 C:\mallet-2.0.8RC2\sample-data\web\en\elizabeth_needham.txt 2 2 died 2
0 C:\mallet-2.0.8RC2\sample-data\web\en\elizabeth_needham.txt 3 3 mother 2
0 C:\mallet-2.0.8RC2\sample-data\web\en\elizabeth_needham.txt 4 1 needham 2
0 C:\mallet-2.0.8RC2\sample-data\web\en\elizabeth_needham.txt 5 4 english 2
0 C:\mallet-2.0.8RC2\sample-data\web\en\elizabeth_needham.txt 6 5 procuress 2
0 C:\mallet-2.0.8RC2\sample-data\web\en\elizabeth_needham.txt 7 6 brothel 2
0 C:\mallet-2.0.8RC2\sample-data\web\en\elizabeth_needham.txt 8 7 keeper 2
0 C:\mallet-2.0.8RC2\sample-data\web\en\elizabeth_needham.txt 9 8 century 2
0 C:\mallet-2.0.8RC2\sample-data\web\en\elizabeth_needham.txt 10 9 london 2
0 C:\mallet-2.0.8RC2\sample-data\web\en\elizabeth_needham.txt 11 10 identified 2
```

图 6-18

- `C:\web.en.keys.txt` 包含着每个主题的编号、权重、热门关键词，这些数据我们在步骤 2 中的控制台输出中已经看到过。
- `C:/web.en.composition.txt` 包含着导入的每个原始文本文件中各个主题的拆解明细。下面内容显示的是这个文件的一部分。如图 6-19 所示，你可以使用任何一种电子表格程序打开它，比如 Microsoft Excel。

#doc name topic proportion ...							
0	file:/C:/m	2	0.998723	10	2.00E-04	13	1.00E-04
1	file:/C:/m	6	0.999228	10	1.12E-04	2	1.08E-04
2	file:/C:/m	7	0.999107	10	1.30E-04	2	1.25E-04
3	file:/C:/m	1	0.827995	10	0.171402	2	9.91E-05
4	file:/C:/m	4	0.999251	10	1.09E-04	2	1.05E-04
5	file:/C:/m	10	0.999527	2	7.18E-05	13	3.73E-05
6	file:/C:/m	12	0.999228	10	1.12E-04	2	1.08E-04
7	file:/C:/m	18	0.999219	10	1.14E-04	2	1.10E-04
8	file:/C:/m	3	0.99943	10	8.31E-05	2	8.00E-05
9	file:/C:/m	14	0.999307	10	1.01E-04	2	9.72E-05
10	file:/C:/m	0	0.972324	2	0.027101	10	9.73E-05
11	file:/C:/m	5	0.704671	13	0.294834	10	7.77E-05

图 6-19

大部分情况下，你都可以使用上面这些关键命令从文章集合中提取主题。上面这些操作步骤适合用来从一系列文本中提取主题。如果你只需要从单个文章中提取主题，请把这个文章放入一个目录下，并且将其视作单文档语料库，采用上面步骤提取主题即可。

最后，让我们了解一下有哪些主题建模算法可以与 Mallet 配合使用。

- LDA
- Parallel LDA
- DMR LDA

- Hierarchical LDA
- Labeled LDA
- Polylingual topic model
- Hierarchical Pachinko allocation model
- Weighted topic model
- LDA with integrated phrase discovery
- 使用 skip-gram 与负采样的 Word Embeddings（word2vec）

6.8 使用 Mallet 对文本文档进行分类

在本章的最后两节，我们将使用传统的机器学习分类问题，即使用语言建模技术对文档进行分类。本节中，我们会使用 Mallet 及其命令行接口训练模型，并把训练好的模型应用到从未见过的测试数据上。

使用 Mallet 进行分类包含如下 3 个步骤：

1. 把训练文档转换成 Mallet 的原生格式；
2. 使用训练文档训练模型；
3. 使用模型对未见过的测试文档进行分类。

其中，"把训练文档转换成 Mallet 的原生格式"从技术上来说是指把文档数据转换成特征向量。我们不需要手动从训练或测试文档中提取任何特征，Mallet 会帮我们做这个工作。你既可以采用物理方法分离训练与测试数据，也可以使用命令行选项获取文档清单、分段训练与测试部分。

让我们设想如下一个简单场景：我们所拥有的文本数据是普通的文本文件，每个文件就是一个文档。我们不需要识别这些文档的开头与结尾。这些文件组织在一系列目录中，所有带有相同类别标签的文档存在于同一个目录下。比如，如果所有文本文件分属于两个类别，即 spam 与 ham，为此我们需要创建两个目录，一个目录用来保存所有 spam 文档，另一个用来保存所有 ham 文档。

准备工作

1. 关于安装 Mallet 的方法，我们已经在上一节中讲解过，这里不再赘述。

2. 打开网页浏览器，在地址栏中输入如下地址：http://www.cs.cmu.edu/afs/cs/project/theo-11/www/naive-bayes/20_newsgroups.tar.gz。下载后的文件夹中包含已经分类好的新闻文章，分别存放在 20 个不同的目录中（见图 6-20）。请把它们复制到 Mallet 安装目录中。

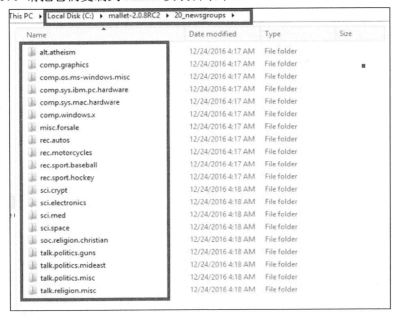

图 6-20

操作步骤

1. 打开命令行窗口，进入到 Mallet 安装文件夹的 bin 目录下。
2. 在 bin 目录下，输入如下命令：

    ```
    mallet import-dir --input C:\mallet-2.0.8RC2\20_newsgroups\*--preserve-case --remove-stopwords --binary-features --gram-sizes 1 --output C:\20_newsgroup.classification.mallet
    ```

执行上面命令，将从 C:\mallet-2.0.8RC2\20_newsgroups 文件夹中取出所有文档，并删除其中的停用词，保留文档中单词实例，使用 gram size 1 创建二元特征（binary features）。如图 6-21 所示，从这些文档输出的 Mallet 原生格式文件存储为 C:\20_newsgroup. classification.mallet。

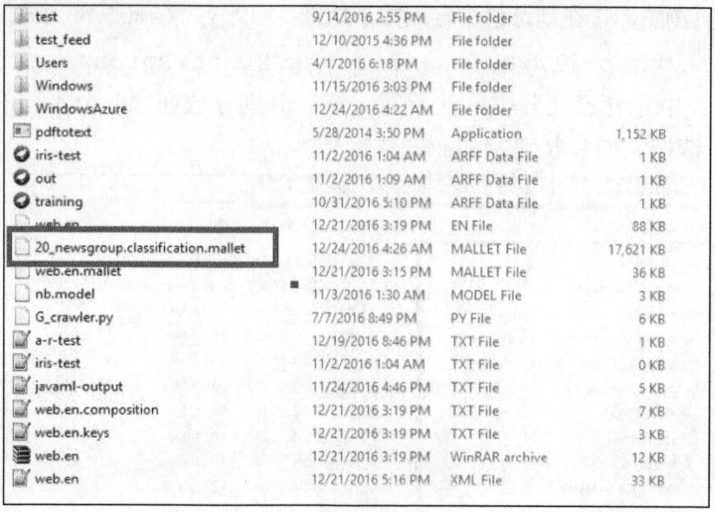

第 6 章 从文本数据提取信息

图 6-21

3. 接着，使用如下命令，根据数据创建最大熵分类器。下面命令把上一步的输出作为输入，使用 1-grams 根据二元特征创建一个朴素贝叶斯分类器，并将其输出为 C:\20_newsgroup.classification.classifier（见图 6-22）。

```
mallet train-classifier --trainer NaiveBayes --input
C:\20_newsgroup.classification.mallet --output-classifier
C:\20_newsgroup.classification.classifier
```

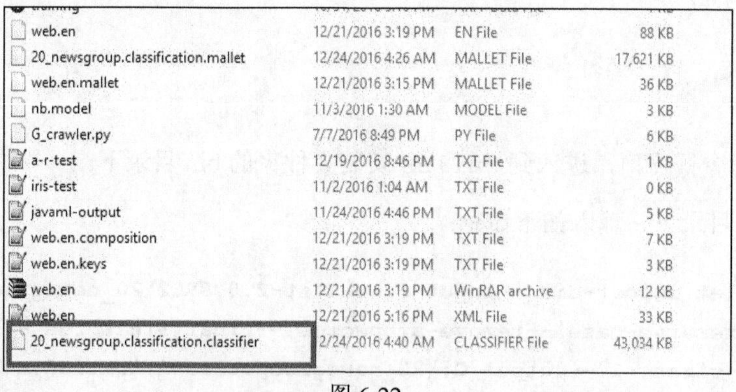

图 6-22

除了朴素贝叶斯之外，Mallet 还支持其他许多算法，完整列表如下。

- AdaBoost
- Bagging

- Winnow
- C45 决策树
- Ensemble trainer
- 最大熵分类器（多项逻辑回归）
- 朴素贝叶斯
- Rank Maximum Entropy Classifier
- Posterior Regularization Auxiliary Model

4. 除了使用整个数据集训练之外，你还可以把一部分数据选出来用作训练数据，把其余数据用作测试数据。基于测试数据的真实类别标签，你可以了解分类器的预测性能。

在命令行窗口中，进入 bin 目录下，输入如下命令：

```
mallet train-classifier --trainer NaiveBayes --input
   C:\20_newsgroup.classification.mallet --training-portion 0.9
```

执行上面命令，将从数据中随机选择 90%，用来训练朴素贝叶斯分类器。最后，把分类器应用到剩余的 10% 的数据上，此时不必查看这些数据的实际标签，只在评估分类器时才考虑这些数据的真实分类，如图 6-23 所示。

```
NaiveBayesTrainer
Summary. train accuracy mean = 0.9678835361449131 stddev = 0.0 stderr = 0.0
Summary. test accuracy mean = 0.915 stddev = 0.0 stderr = 0.0
Summary. test precision(alt.atheism) mean = 0.7272727272727273 stddev = 0.0 stde
rr = 0.0
Summary. test precision(comp.graphics) mean = 0.89 stddev = 0.0 stderr = 0.0
Summary. test precision(comp.os.ms-windows.misc) mean = 0.975609756097561 stddev
 = 0.0 stderr = 0.0
Summary. test precision(comp.sys.ibm.pc.hardware) mean = 0.9019607843137255 stdd
ev = 0.0 stderr = 0.0
Summary. test precision(comp.sys.mac.hardware) mean = 0.9230769230769231 stddev
= 0.0 stderr = 0.0
Summary. test precision(comp.windows.x) mean = 0.968421052631579 stddev = 0.0 st
derr = 0.0
Summary. test precision(misc.forsale) mean = 0.9175257731958762 stddev = 0.0 std
err = 0.0
Summary. test precision(rec.autos) mean = 0.9702970297029703 stddev = 0.0 stderr
 = 0.0
Summary. test precision(rec.motorcycles) mean = 0.989010989010989 stddev = 0.0 s
tderr = 0.0
Summary. test precision(rec.sport.baseball) mean = 1.0 stddev = 0.0 stderr = 0.0
Summary. test precision(rec.sport.hockey) mean = 1.0 stddev = 0.0 stderr = 0.0
Summary. test precision(sci.crypt) mean = 0.9897959183673469 stddev = 0.0 stderr
 = 0.0
Summary. test precision(sci.electronics) mean = 0.9042553191489362 stddev = 0.0
stderr = 0.0
Summary. test precision(sci.med) mean = 1.0 stddev = 0.0 stderr = 0.0
Summary. test precision(sci.space) mean = 0.9711538461538461 stddev = 0.0 stderr
```

图 6-23

这个命令给出了分类器对 20 个分类的总准确度，包括每个类的准确率、召回率、精确度以及标准误差。

5. 你也可以多次进行训练与测试，每次的训练集与测试集都是随机选择的。比如，如果你打算使用 90%的数据来训练分类器，使用余下的 10%来测试分类器，重复 10 次，并且采用随机方式划分数据集，请使用如下命令：

```
mallet train-classifier --trainer NaiveBayes --input
C:\20_newsgroup.classification.mallet --training-portion 0.9--
num-trials 10
```

6. 此外，我们还可以使用 Mallet 进行交叉验证，在这一过程中，我们可以指定交叉验证期间要创建的折数。比如，如果你打算进行 10 折交叉验证，请使用如下命令：

```
mallet train-classifier --trainer NaiveBayes --input
C:\20_newsgroup.classification.mallet --cross-validation 10
```

执行上面命令，将给出 10 次试验中每一次的结果（见图 6-24），每一次所使用的测试数据都是从源数据中重新获取的，同时还会给出 10 次试验的平均结果。Mallet 还提供混淆矩阵，这对于数据科学家是非常重要的，因为可以帮助他们更好地理解模型。

```
Trial 9 Training NaiveBayesTrainer with 17997 instances
Trial 9 Training NaiveBayesTrainer finished
Trial 9 Trainer NaiveBayesTrainer training data accuracy = 0.9671611935322554
Trial 9 Trainer NaiveBayesTrainer Test Data Confusion Matrix
Confusion Matrix, row=true, column=predicted  accuracy=0.917 most-frequent-tag b
aseline=0.0595
                      label   0    1   2   3   4   5   6   7   8   9   10  11  12
 13  14  15  16  17  18  19 |total
 0              alt.atheism  73    .   .   .   .   .   .   .   .   .   .   .   .
  .   .   .   .   .   .  24 |97
 1             comp.graphics   .  100   .   .   .   1   1   .   .   .   .   .   1
  .   1   .   .   1   .   . |105
 2    comp.os.ms-windows.misc   .   4  83  16   .   .   .   .   1   .   .   .   .
  .   .   .   .   .   .   . |105
 3    comp.sys.ibm.pc.hardware   .   1   2  91   3   .   2   .   .   .   .   .   2
  .   .   .   .   .   .   . |101
 4      comp.sys.mac.hardware   .   .   1   .  92   .   .   .   .   .   .   .   1
  .   .   .   .   .   .   . |94
 5             comp.windows.x   .   2   .   2   .  96   .   .   .   .   .   .   .
  .   .   .   .   .   .   . |100
 6              misc.forsale   .   .   .   .   2   1   .  92   2   .   .   .   .
  .   1   1   1   .   .   . |100
 7                 rec.autos   .   .   .   .   .   .   1   .   3 102   .   .   .
  .   .   .   .   .   .   . |106
 8           rec.motorcycles   .   .   .   .   .   .   .   .   .   2 114   .   .
  .   1   .   2   .   .   . |119
 9         rec.sport.baseball   .   .   .   .   .   .   .   .   .   .   1   1 100
  1   .   .   .   .   .   . |103
 10          rec.sport.hockey   .   .   .   .   .   .   .   .   .   .   .   .   .
  .   .   .   .  96   .   . |97
 11                 sci.crypt   .   .   .   .   .   .   .   .   .   .   .   .   .
  .   .   .   .   . 103   . |105
 12           sci.electronics   .   .   .   .   .   .   2   .   .   .   .   .   .
  .   .   .   .   .   1 . 112
```

图 6-24

7. Mallet 还允许我们比较使用不同算法开发的多种分类器的性能。比如，如下命令将使用 10 折交叉验证对两个分类器，即朴素贝叶斯分类器与最大熵分类器进行比较。

```
mallet train-classifier --trainer MaxEnt --trainer NaiveBayes-
  -input C:\20_newsgroup.classification.mallet --cross-
    validation 10
```

8. 如果你想对一些未见过的测试文档（不是指本例，因为在第二步中我们已经把整个文档目录用来训练模型了）应用已经保存过的分类器，你可以使用如下命令。

```
mallet classify-dir --input <directory containing unseen test
  data> --output - --classifier
    C:\20_newsgroup.classification.classifier
```

执行上面命令后，将在控制台中显示对未见过的测试文档的预测分类。此外，你还可以使用如下命令把预测结果保存在 TSV 文件中。

```
mallet classify-dir --input <directory containing unseen test
  data> --output <Your output file> --classifier
    C:\20_newsgroup.classification.classifier
```

9. 最后，我们还可以向单个测试文档（这些文档之前未见过）应用已经保存的分类器。为此，你可以使用如下命令。

```
mallet classify-file --input <unseen test data file path> --
output - --classifier
  C:\20_newsgroup.classification.classifier
```

执行上面命令后，将在控制台中显示对未见过的测试文档的预测分类。此外，你还可以使用如下命令把预测结果保存在 TSV 文件中。

```
mallet classify-file --input <unseen test data file path> --
output <Your output file> --classifier C:\20_ne
  wsgroup.classification.classifier
```

6.9　使用 Weka 对文本文档进行分类

在第 4 章中，我们已经学习过如何使用 Weka 对非文本格式的数据点进行分类。在使用机器学习模型对文本文档进行分类时，Weka 也是一个非常有用的工具。本节中，我们将演示如何使用 Weka 3 开发文档分类模型。

准备工作

1. 下载 Weka。在下载页面中，你会看到针对 Windows、Mac OS X 以及其他操作系统平台（比如 Linux）的下载链接。请认真阅读下载说明，选择相应版本的 Weka 进行下载。

   写作本书之时，面向开发者的 Weka 最新版本为 3.9.0。由于我使用的是 64 位的 Windows 操作系统，并且已经安装好了 JVM 1.8，所以在 Window 版块中选择不带 Java VM 的自解压可执行文件进行下载。

2. 下载完成后，双击可执行文件，根据屏幕提示进行安装即可。请注意，要选择 Weka 的完整版本进行安装。

3. 安装完成后，在运行软件之前，先去安装目录，找到 Weka 对应的 Java 文档文件（weka.java），将其作为外部库添加到你的 Eclipse 项目中。

4. 本节中要使用的示例文档文件保存在相应的目录下，并且每个目录包含同类别的文档。为了下载示例文档，打开网页浏览器，在地址栏中输入如下地址：https://weka.wikispaces.com/file/view/text_example.zip/82917283/text_example.zip。此时，浏览器弹出对话框询问文件保存的位置（只有当浏览器配置为询问保存位置才会弹出这个对话框），把文件保存到 C:/目录下。下载完成后，进行解压缩，得到如图 6-25 所示的目录结构。

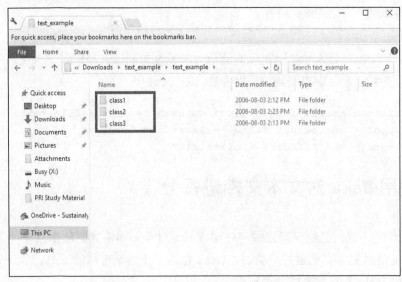

图 6-25

每个目录中都包含几个 html 文件,它们属性某个特定的类,这些类的标签分别为 class1、class2、class3。

至此,我们为使用 Weka 对文档进行分类的所有准备工作就完成了。

操作步骤

1. 创建名为 WekaClassification 的类,并在其中创建 main() 方法,所有代码都将放入该方法中。main() 方法可能会抛出异常。

   ```
   public class WekaClassification {
     public static void main(String[] args) throws Exception {
   ```

2. 创建一个加载器,并设置父目录的路径,以加载所有类目录的父目录。

   ```
   TextDirectoryLoader loader = new TextDirectoryLoader();
   loader.setDirectory(new File("C:/text_example"));
   ```

3. 使用加载的 html 文件创建实例。

   ```
   Instances data = loader.getDataSet();
   ```

4. 使用数据字符串创建词向量。为此,首先创建一个过滤器,用来把字符串转换为词向量,然后把上一步得到的原数据设置给它。

   ```
   StringToWordVector filter = new StringToWordVector();
   filter.setInputFormat(data);
   ```

5. 为了完成字符串到词向量的转换,使用这个过滤器与数据创建实例。

   ```
   Instances dataFiltered = Filter.useFilter(data, filter);
   ```

6. 使用词向量创建朴素贝叶斯分类器。

   ```
   NaiveBayes nb = new NaiveBayes();
   nb.buildClassifier(dataFiltered);
   ```

7. 此时,你可能想看看模型的样子。为此,只需使用如下语句,将模型在控制台中打印出来即可。

   ```
   System.out.println("\n\nClassifier model:\n\n" + nb);
   ```

8. 模型的部分输出如图 6-26 所示。

```
smashed
    mean                                                     0       0 0.3333
    std. dev.                                           0.1667  0.1667 0.4714
    weight sum                                               3       1      3
    precision                                                1       1      1
social
    mean                                                     0       0 0.3333
    std. dev.                                           0.1667  0.1667 0.4714
    weight sum                                               3       1      3
    precision                                                1       1      1
solely
    mean                                                     0       0 0.3333
    std. dev.                                           0.1667  0.1667 0.4714
    weight sum                                               3       1      3
    precision                                                1       1      1
```

图 6-26

9. 接下来，我们使用 K 折交叉验证来评估模型，编写如下代码。

```
Evaluation eval = null;
eval = new Evaluation(dataFiltered);
eval.crossValidateModel(nb, dataFiltered, 5, new Random(1));
System.out.println(eval.toSummaryString());
```

执行上面代码，在控制台中打印出对分类器的评估结果。

```
Correctly Classified Instances           1               14.2857 %
Incorrectly Classified Instances         6               85.7143 %
Kappa statistic                         -0.5
Mean absolute error                      0.5714
Root mean squared error                  0.7559
Relative absolute error                126.3158 %
Root relative squared error            153.7844 %
Total Number of Instances                7
```

请注意，上面代码中我们使用了 5 折交叉验证，并未使用标准的 10 折交叉验证，因为文档总数少于 10 个（准确地说是 7 个）。

示例完整代码整理如下：

```
import weka.core.*;
import weka.core.converters.*;
import weka.classifiers.Evaluation;
import weka.classifiers.bayes.NaiveBayes;
import weka.filters.*;
import weka.filters.unsupervised.attribute.*;
```

```java
import java.io.*;
import java.util.Random;

public class WekaClassification {
    public static void main(String[] args) throws Exception {
        TextDirectoryLoader loader = new TextDirectoryLoader();
        loader.setDirectory(new File("C:/text_example"));
        Instances data = loader.getDataSet();

        StringToWordVector filter = new StringToWordVector();
        filter.setInputFormat(data);
        Instances dataFiltered = Filter.useFilter(data, filter);

        NaiveBayes nb = new NaiveBayes();
        nb.buildClassifier(dataFiltered);
        System.out.println("\n\nClassifier model:\n\n" + nb);
        Evaluation eval = null;
        eval = new Evaluation(dataFiltered);
        eval.crossValidateModel(nb, dataFiltered, 5, new Random(1));
        System.out.println(eval.toSummaryString());
    }
}
```

第 7 章 处理大数据

本章涵盖如下内容：

- 使用 Apache Mahout 训练在线逻辑回归模型；
- 使用 Apache Mahout 应用在线逻辑回归模型；
- 使用 Apache Spark 解决简单文本挖掘问题；
- 使用 MLib 的 K 均值算法进行聚类；
- 使用 MLib 创建线性回归模型；
- 使用 MLib 的随机森林模型对数据点进行分类。

7.1 简介

本章，我们将学习大数据框架中使用的 3 种关键技术，分别是 Apache Mahout、Apache Spark，及其机器学习库 MLib，它们对于数据科学家极其有用。

首先学习 Apache Mahout，它是一个可扩展的、分布式机器学习平台，可以用来进行分类、回归、聚类、协同过滤任务。Mahout 起初是一个机器学习工作台，只工作在 Hadoop MapReduce 之上，但是最后选择 Apache Spark 作为它的平台。

Apache Spark 是一个支持大数据并行处理的框架，与 MapReduce 很相似，它也支持数据的跨集群分发。但是 Spark 与 MapReduce 最主要的不同在于，Spark 会优先考虑并尽量把数据保存在内存中，而 MapReduce 会不断地读写磁盘。因此，Spark 运行速度要明显快于 MapReduce。下面我们将学习作为一个数据科学家如何使用 Sark 来做简单的文本挖掘任务，比如统计空行数量，获取特定单词在一个大文件中出现的频率。选用 Spark 的另一个原因是它不仅支持 Java，还支持其他流行语言，比如 Python、Scala，而使用 MapReduce

时，通常只能选用 Java 语言。

MLib 是一个可扩展的机器学习库，它来自于 Apache Spark，包含各种分类、回归、聚类、协同过滤、特征选择算法。MLib 通常工作在 Spark 下，借助其速度来解决机器学习问题。本章，我们将学习如何使用这个库来解决分类、回归、聚类问题。

 本书中我们使用的是 Mahout 0.9 版本，感兴趣的读者可以进一步了解 Mahout 0.10.x 与 MLib 之间的不同。

7.2　使用 Apache Mahout 训练在线逻辑回归模型

本节，我们将学习使用 Apache Mahout Java 库来训练在线逻辑回归模型。

准备工作

1. 首先要在 Eclipse 中新建一个 Maven 项目。我选用的是 Eclipse Mars，如图 7-1 所示，为了新建 Maven 项目，依次选择"File|New|Other..."菜单。

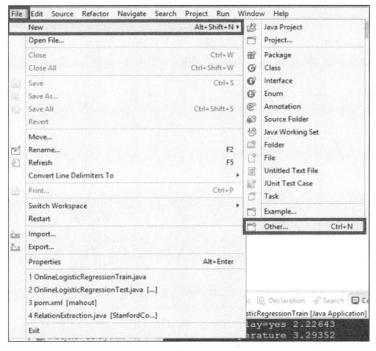

图 7-1

2. 然后，在向导中展开 Maven，选择 Maven 项目。不断单击 Next，直到 Eclipse 提示你输入 Artifact Id（见图 7-2），输入 `mahout`，此时灰色 Finish 按钮会被激活，单击 Finish 按钮，创建好一个名为 mahout 的 Maven 项目。

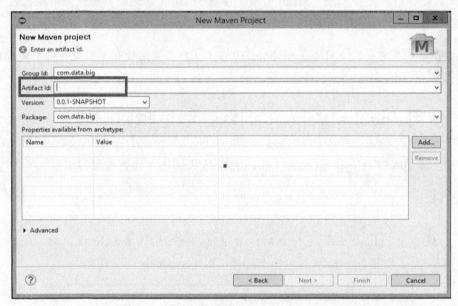

图 7-2

3. 如图 7-3 所示，在 Eclipse 的包浏览器中，双击 `pom.xml` 文件编辑它。

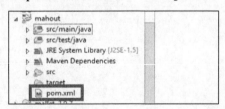

图 7-3

4. 单击 `pom.xml` 选项卡，显示出 `pom.xml` 文件的内容，把如下代码行添加到 `pom.xml` 文件的`<dependencies>...</dependencies>`标记之间，保存文件修改。这样会自动地把依赖的 JAR 文件下载到项目中。

```
<dependency>
    <groupId>org.apache.mahout</groupId>
    <artifactId>mahout-core</artifactId>
    <version>0.9</version>
</dependency>
<dependency>
```

```xml
        <groupId>org.apache.mahout</groupId>
        <artifactId>mahout-examples</artifactId>
        <version>0.9</version>
    </dependency>
    <dependency>
        <groupId>org.apache.mahout</groupId>
        <artifactId>mahout-math</artifactId>
        <version>0.9</version>
    </dependency>
```

5. 如图 7-4 所示，在 mahout 项目的 `src/main/java` 目录下，创建一个名为 `chap7.science.data` 的包。

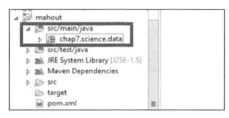

图 7-4

6. 在项目名称上右击，选择 **New**，再选择文件夹。创建两个文件夹，第一个文件夹的名称为 `data`，用来存放输入数据集，我们将为它创建模型；第二个文件夹的名称为 `model`，用来保存模型。输入 `data` 作为文件夹名称，单击 **Finish**，重复这个步骤，创建另一个名为 `model` 的文件夹。

7. 在 `data` 文件夹下，创建一个名称为 `weather.numeric.csv` 的 CSV 文件，里面包含如下数据。

```
outlook,temperature,humidity,windy,play
sunny,85,85,FALSE,no
sunny,80,90,TRUE,no
overcast,83,86,FALSE,yes
rainy,70,96,FALSE,yes
rainy,68,80,FALSE,yes
rainy,65,70,TRUE,no
overcast,64,65,TRUE,yes
sunny,72,95,FALSE,no
sunny,69,70,FALSE,yes
rainy,75,80,FALSE,yes
sunny,75,70,TRUE,yes
overcast,72,90,TRUE,yes
overcast,81,75,FALSE,yes
rainy,71,91,TRUE,no
```

8. 接下来，要开始编写代码了。

操作步骤

1. 在刚创建的包中，创建一个名为 `OnlineLogisticRegressionTrain.java` 的 Java 类文件。双击该类文件，在其中编写代码，创建一个名称为 `OnlineLogisticRegressionTrain` 的类。

    ```
    public class OnlineLogisticRegressionTrain {
    ```

2. 编写 `main()` 方法如下：

    ```
    public static void main(String[] args) throws IOException {
    ```

3. 创建两个 `String` 类型的变量，分别用来存放输入数据文件路径与模型文件（我们要创建与保存的模型）路径。

    ```
    String inputFile = "data/weather.numeric.csv";
    String outputFile = "model/model";
    ```

4. 接着，创建一个用来包含数据文件特征的列表。

    ```
    List<String> features =Arrays.asList("outlook", "temperature",
        "humidity", "windy", "play");
    ```

5. 在这一步，我们把数据文件中的所有特征名称列出来，保持它们在数据文件中的出现顺序不变。

6. 接下来，定义每个特征的类型。特征类型字符 w 表示一个名义特征，n 表示数值特征。

    ```
    List<String> featureType = Arrays.asList("w", "n", "n", "w",
        "w");
    ```

7. 现在，我们要为分类器设置参数。在这一步，我们将创建一个参数变量，并且设置几个值作为参数。我们会设置目标变量或类变量（本例中是 play）。观察一下数据，你会发现类变量 play 最多有两个值，即 yes 或 no。因此，我们把最大目标类别设置为 2。接着，设置特征数量（本例中是 4），注意并不是类特征。接下来的 3 个参数取决于算法。本节中，在生成分类器时我们不会使用任何偏差率（bias），而使用一个平衡的学习率 0.5。最后，使用类型图方法设置特征及其类型。

```
LogisticModelParameters params = new
   LogisticModelParameters();
params.setTargetVariable("play");
params.setMaxTargetCategories(2);
params.setNumFeatures(4);
params.setUseBias(false);
params.setTypeMap(features,featureType);
params.setLearningRate(0.5);
```

8. 接下来，使用 10 passes 创建分类器。这个数字是随意设定的，你根据自己的经验设置一个合适的值。

    ```
    int passes = 10;
    ```

9. 创建在线线性回归分类器。

    ```
    OnlineLogisticRegression olr;
    ```

10. 创建一个变量，用来从 CSV 文件读取数据，然后开始创建回归模型。

    ```
    CsvRecordFactory csv = params.getCsvRecordFactory();
    olr = params.createRegression();
    ```

11. 接着，创建一个 for 循环，用来遍历 10 个 passes。

    ```
    for (int pass = 0; pass < passes; pass++) {
    ```

12. 开始读取数据文件。

    ```
    BufferedReader in = new BufferedReader(new
       FileReader(inputFile));
    ```

13. 获取数据文件的头部，它由一系列特征名称组成。

    ```
    csv.firstLine(in.readLine());
    ```

14. 读取数据行。

    ```
    String row = in.readLine();
    ```

15. 接着遍历非 null 的每一行。

    ```
    while (row != null) {
    ```

16. 为每一行（或数据行）显示数据点，创建一个输入向量。

    ```
    System.out.println(row);
    Vector input = new
      RandomAccessSparseVector(params.getNumFeatures());
    ```

17. 为数据行获取 `targetValue`。

    ```
    int targetValue = csv.processLine(row, input);
    ```

18. 使用数据点训练模型。

    ```
    olr.train(targetValue, input);
    ```

19. 读取下一行。

    ```
    row = in.readLine();
    ```

20. 关闭循环。

    ```
    }
    ```

21. 关闭读取输入数据文件。

    ```
    in.close();
    ```

22. 关闭遍历 `passes` 的循环。

    ```
    }
    ```

23. 最后，把 `output` 模型保存到名为 `model` 的文件中，该文件位于 Eclipse 项目的 `model` 目录中。

    ```
    OutputStream modelOutput = new FileOutputStream(outputFile);
    try {
        params.saveTo(modelOutput);
    } finally {
        modelOutput.close();
    }
    ```

24. 关闭 `main` 方法与类。

    ```
    }
    ```

    }

25. 运行上面代码之后，在控制台的输出中，你会看到数据文件中的数据行。并且学习好的模型会被保存到项目的 model 目录中。

示例的完整代码整理如下：

```java
package chap7.science.data;

import java.io.BufferedReader;
import java.io.FileOutputStream;
import java.io.FileReader;
import java.io.IOException;
import java.io.OutputStream;
import java.util.Arrays;
import java.util.List;
import org.apache.mahout.classifier.sgd.CsvRecordFactory;
import org.apache.mahout.classifier.sgd.LogisticModelParameters;
import org.apache.mahout.classifier.sgd.OnlineLogisticRegression;
import org.apache.mahout.math.RandomAccessSparseVector;
import org.apache.mahout.math.Vector;

public class OnlineLogisticRegressionTrain {
    public static void main(String[] args) throws IOException {
        String inputFile = "data/weather.numeric.csv";
        String outputFile = "model/model";

        List<String> features =Arrays.asList("outlook", "temperature",
           "humidity", "windy", "play");
        List<String> featureType = Arrays.asList("w", "n", "n", "w",
           "w");
        LogisticModelParameters params = new LogisticModelParameters();
        params.setTargetVariable("play");
        params.setMaxTargetCategories(2);
        params.setNumFeatures(4);
        params.setUseBias(false);
        params.setTypeMap(features,featureType);
        params.setLearningRate(0.5);

        int passes = 10;
        OnlineLogisticRegression olr;

        CsvRecordFactory csv = params.getCsvRecordFactory();
```

```
    olr = params.createRegression();

    for (int pass = 0; pass < passes; pass++) {
      BufferedReader in = new BufferedReader(new
        FileReader(inputFile));
      csv.firstLine(in.readLine());
      String row = in.readLine();
      while (row != null) {
        System.out.println(row);
        Vector input = new
          RandomAccessSparseVector(params.getNumFeatures());
        int targetValue = csv.processLine(row, input);
        olr.train(targetValue, input);
        row = in.readLine();
      }
      in.close();
    }

    OutputStream modelOutput = new FileOutputStream(outputFile);
    try {
      params.saveTo(modelOutput);
    } finally {
      modelOutput.close();
    }
  }
}
```

7.3 使用 Apache Mahout 应用在线逻辑回归模型

本节中，我们将演示如何使用 Apache Mahout 把在线逻辑回归模型应用到从未见过且未打标签的测试数据上。请注意，本节内容与上一节内容联系紧密，要求我们先使用训练数据创建一个模型，关于模型的创建方法，请参考上一节的内容。

准备工作

1. 参照上一节内容，创建好模型之后，进入项目文件夹，找到上一节中创建的名为 `model` 的目录，你应该能够看到其中包含一个 `model` 文件。

2. 接着，创建测试文件。在上一节中，我们已经在项目文件夹中创建了一个名为 `data` 的文件夹，进入该文件夹，创建一个名为 `weather.numeric.test.csv` 的测试文件，里面包含如下数据：

```
outlook,temperature,humidity,windy,play
overcast,90,80,TRUE,yes
overcast,95,88,FALSE,yes
rainy,67,78,TRUE,no
rainy,90,97,FALSE,no
sunny,50,67,FALSE,yes
sunny,67,75,TRUE,no
```

3. 在名为 mahout 的 Eclipse 项目中，在 src/main/java 文件夹中，应该有一个名为 chap7.science.data 的包，该文件夹是我们在上一节中创建的。在这个包中，创建一个名为 OnlineLogisticRegressionTest.java 的 Java 类文件，双击该文件，将其打开。

操作步骤

1. 创建 OnlineLogisticRegressionTest 类。

    ```
    public class OnlineLogisticRegressionTest {
    ```

2. 接下来，声明几个类变量。首先，创建两个变量，用来存放 data 文件与 model 文件的路径（这两个文件我们已经在上一节中创建好了）。

    ```
    private static String inputFile =
       "data/weather.numeric.test.csv";
    private static String modelFile = "model/model";
    ```

3. 然后开始编写 main 方法。

    ```
    public static void main(String[] args) throws Exception {
    ```

4. 创建一个 AUC 类型的变量，计算分类器的 AUC（曲线下面积），并且把它用作性能指标。

    ```
    Auc auc = new Auc();
    ```

5. 接下来，从 model 文件读取并加载在线逻辑回归算法的参数。

    ```
    LogisticModelParameters params =
       LogisticModelParameters.loadFrom(new File(modelFile));
    ```

6. 创建一个变量，以便读取测试数据文件。

    ```
    CsvRecordFactory csv = params.getCsvRecordFactory();
    ```

7. 创建一个 onlinelogisticregression 分类器。

    ```
    OnlineLogisticRegression olr = params.createRegression();
    ```

8. 读取测试数据文件。

    ```
    InputStream in = new FileInputStream(new File(inputFile));
    BufferedReader reader = new BufferedReader(new
      InputStreamReader(in, Charsets.UTF_8));
    ```

9. 测试数据文件的第一行是文件头或特征类别。因此，需要从分类中忽略这一行，从下一行（或数据点）开始读取。

    ```
    String line = reader.readLine();
    csv.firstLine(line);
    line = reader.readLine();
    ```

10. 你可能想把分类结果显示在控制台中。为此，我们创建一个 PrintWriter 变量。

    ```
    PrintWriter output=new PrintWriter(new
      OutputStreamWriter(System.out, Charsets.UTF_8), true);
    ```

11. 我们将把预测类别、模型输出以及对数似然值打印出来。创建头部，把它们打印在控制台中。

    ```
    output.println(""class","model-output","log-likelihood"");
    ```

12. 接着，遍历每个非 null 行。

    ```
    while (line != null) {
    ```

13. 为测试数据创建特征向量。

    ```
    Vector vector = new
      SequentialAccessSparseVector(params.getNumFeatures());
    ```

14. 创建一个变量，用来存放每一行或每个数据点的真实类别值。

```
            int classValue = csv.processLine(line, vector);
```

15. 对测试数据点进行分类,并从分类器获取分数。

    ```
            double score = olr.classifyScalarNoLink(vector);
    ```

16. 使用如下代码把类值、分数、对数似然值打印到控制台中。

    ```
            output.printf(Locale.ENGLISH, "%d,%.3f,%.6f%n", classValue,
               score, olr.logLikelihood(classValue, vector));
    ```

17. 把 score 与 classvalue 添加到 AUC 变量中。

    ```
            auc.add(classValue, score);
    ```

18. 读取下一行,关闭循环。

    ```
            line = reader.readLine();
         }
    ```

19. 关闭 reader。

    ```
         reader.close();
    ```

20. 接下来,让我们把分类结果打印出来。首先,打印 AUC。

    ```
         output.printf(Locale.ENGLISH, "AUC = %.2f%n", auc.auc());
    ```

21. 接着,打印分类的混淆矩阵。为此,先创建一个混淆矩阵。由于训练与测试数据只有两个类,所以我们会得到一个 2×2 的混淆矩阵。

    ```
         Matrix matrix = auc.confusion();
         output.printf(Locale.ENGLISH, "confusion: [[%.1f, %.1f], [%.1f,
            %.1f]]%n", matrix.get(0, 0), matrix.get(1, 0), matrix.get(0,
            1), matrix.get(1, 1));
    ```

22. matrix 中存放着熵值。虽然我们不必为此新建一个 matrix 变量,但是只要你愿意,你仍然可以这样做。

    ```
         matrix = auc.entropy();
         output.printf(Locale.ENGLISH, "entropy: [[%.1f, %.1f], [%.1f,
            %.1f]]%n", matrix.get(0, 0), matrix.get(1, 0), matrix.get(0,
            1), matrix.get(1, 1));
    ```

23. 关闭 main 方法与类。

            }
        }

示例的完整代码整理如下:

```java
package chap7.science.data;

import com.google.common.base.Charsets;
import org.apache.mahout.math.Matrix;
import org.apache.mahout.math.SequentialAccessSparseVector;
import org.apache.mahout.math.Vector;
import org.apache.mahout.classifier.evaluation.Auc;
import org.apache.mahout.classifier.sgd.CsvRecordFactory;
import org.apache.mahout.classifier.sgd.LogisticModelParameters;
import org.apache.mahout.classifier.sgd.OnlineLogisticRegression;
import java.io.BufferedReader;
import java.io.File;
import java.io.FileInputStream;
import java.io.InputStream;
import java.io.InputStreamReader;
import java.io.OutputStreamWriter;
import java.io.PrintWriter;
import java.util.Locale;

public class OnlineLogisticRegressionTest {

    private static String inputFile = "data/weather.numeric.test.csv";
    private static String modelFile = "model/model";

    public static void main(String[] args) throws Exception {
        Auc auc = new Auc();
        LogisticModelParameters params =
            LogisticModelParameters.loadFrom(new File(modelFile));
        CsvRecordFactory csv = params.getCsvRecordFactory();
        OnlineLogisticRegression olr = params.createRegression();
        InputStream in = new FileInputStream(new File(inputFile));
        BufferedReader reader = new BufferedReader(new
            InputStreamReader(in, Charsets.UTF_8));
        String line = reader.readLine();
        csv.firstLine(line);
        line = reader.readLine();
        PrintWriter output=new PrintWriter(new
            OutputStreamWriter(System.out, Charsets.UTF_8), true);
```

```
        output.println("\"class\",\"model-output\",\"log-likelihood\"");
        while (line != null) {
            Vector vector = new
                SequentialAccessSparseVector(params.getNumFeatures());
            int classValue = csv.processLine(line, vector);
            double score = olr.classifyScalarNoLink(vector);
            output.printf(Locale.ENGLISH, "%d,%.3f,%.6f%n", classValue,
                score, olr.logLikelihood(classValue, vector));
            auc.add(classValue, score);
            line = reader.readLine();
        }
        reader.close();
        output.printf(Locale.ENGLISH, "AUC = %.2f%n", auc.auc());
        Matrix matrix = auc.confusion();
        output.printf(Locale.ENGLISH, "confusion: [[%.1f, %.1f], [%.1f,
           %.1f]]%n", matrix.get(0, 0), matrix.get(1, 0), matrix.get(0,
             1), matrix.get(1, 1));
        matrix = auc.entropy();
        output.printf(Locale.ENGLISH, "entropy: [[%.1f, %.1f], [%.1f,
           %.1f]]%n", matrix.get(0, 0), matrix.get(1, 0), matrix.get(0,
             1), matrix.get(1, 1));
    }
}
```

运行上面代码，得到如下输出：

```
"class","model-output","log-likelihood"
1,119.133,0.000000
1,123.028,0.000000
0,15.888,-15.887942
0,63.213,-100.000000
1,-6.692,-6.693089
0,24.286,-24.286465
AUC = 0.67
confusion: [[0.0, 1.0], [3.0, 2.0]]
entropy: [[NaN, NaN], [0.0, -9.2]]
```

7.4　使用 Apache Spark 解决简单的文本挖掘问题

根据 Apache Spark 的官网介绍，在内存中 Spark 运行程序的速度比 Hadoop MapReduce 快

100 倍，在磁盘上快 10 倍。一般来说，Apache Spark 是一个开源的集群计算框架，其处理引擎运行速度更快，更易使用，并为数据科学家提供复杂精密的分析功能。

本节中，我们将演示如何使用 Apache Spark 来解决非常简单的数据问题。当然，这里我们用作示例的数据问题只是虚拟的，并非真实的问题，但是它仍能帮助我们直观地了解如何大规模地使用 Apache Spark 这个框架。

准备工作

1. 首先要在 Eclipse 中新建一个 Maven 项目。我使用的是 Eclipse Mars，为了新建 Maven 项目，如图 7-5 所示，在菜单栏中，依次选择"**File|New|Other...**"菜单。

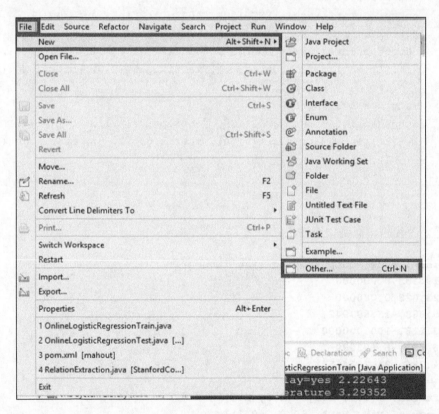

图 7-5

2. 然后，在向导中展开 **Maven**，选择 Maven 项目。如图 7-6 所示，不断单击 **Next**，直到 Eclipse 提示你输入 **Artifact Id** 时，请输入 mlib，此时灰色 Finish 按钮会被激活，单击 **Finish** 按钮，创建好一个名为 mlib 的 Maven 项目。

7.4 使用 Apache Spark 解决简单的文本挖掘问题

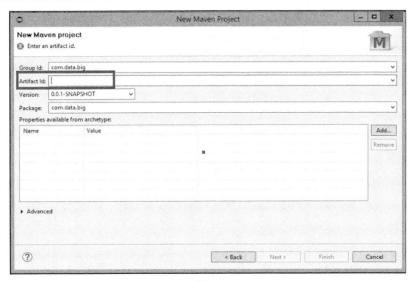

图 7-6

3. 如图 7-7 所示,在 Eclipse 的包浏览器中,双击 pom.xml 文件编辑它。

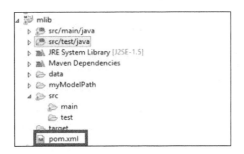

图 7-7

4. 单击 pom.xml 选项卡,显示出 pom.xml 文件的内容,把如下代码行添加到 pom.xml 文件的<dependencies>...</dependencies>标记之间,保存文件修改。这样会自动地把依赖的 JAR 文件下载到项目中。

```
<dependency>
  <groupId>org.apache.spark</groupId>
  <artifactId>spark-mllib_2.10</artifactId>
  <version>1.3.1</version>
</dependency>
```

5. 如图 7-8 所示,在 mlib 项目的 src/main/java 目录下,创建一个名为 com.data.big.mlib 的包。

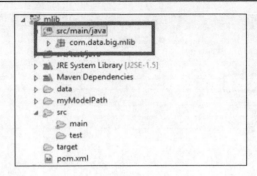

图 7-8

6. 在项目名称上右击，选择 **New**，再选择文件夹。创建一个名为 `data` 的文件夹，用来存放本节要用到的所有数据文件。

7. 本节我们会用到威廉·莎士比亚的著作（文本格式）。打开一个网页浏览器，前往 `http://norvig.com/ngrams/`，你将打开一个名为 Natural Language Corpus Data: Beautiful Data 的页面。如图 7-9 所示，在 Files for Download 版块中，你会看到一个名为 `shakespeare` 的 .txt 文件，将其下载到你的电脑中。

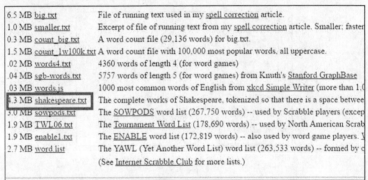

图 7-9

8. 在上面创建的包中，创建一个名为 `SparkTest` 的 `Java` 类文件。双击它将其打开，准备编写代码。

操作步骤

1. 创建 SparkTest 类。

   ```
   public class SparkTest {
   ```

2. 开始编写 `main()` 方法。

```
    public static void main( String[] args ){
```

3. 首先，获取输入数据文件的路径。这里的输入文件是指我们前面下载的莎士比亚的作品，它存在于项目的 data 文件夹中。

```
    String inputFile = "data/shakespeare.txt";
```

4. Spark 的属性用来控制应用程序设置，每个应用程序都要单独进行配置。设置这些属性时，我们可以使用传递给 SparkContext 的 SparkConf。通过 SparkConf，我们可以配置一些通用属性。

```
    SparkConf configuration = new
       SparkConf().setMaster("local[4]").setAppName("My App");
    JavaSparkContext sparkContext = new
       JavaSparkContext(configuration);
```

5. 请注意，在上面代码中，如果我们使用了 local[2]，程序将进入最小的并行执行状态。而在上面代码中我们使用了 local[4]，这表示允许应用程序并行运行 4 个线程。

6. JavaRDD 是一个分布式的对象集合。创建一个 RDD 对象，用来收集 shakespeare.txt 文件中的空行。

```
    JavaRDD<String> rdd =
       sparkContext.textFile(inputFile).cache();
```

 如果使用 local[*]，Spark 将使用系统中的所有内核。

7. 统计输入数据文件中所包含的空行数。

```
    long emptyLines = rdd.filter(new Function<String,Boolean>(){
      private static final long serialVersionUID = 1L;
      public Boolean call(String s){
      return s.length() == 0;
      }
    }).count();
```

8. 把文件中的空行数（emptylines）打印在控制台中。

```
    System.out.println("Empty Lines: " + emptyLines);
```

9. 接着，编写如下代码片段，从输入数据文件获取词频。

```
    JavaPairRDD<String, Integer> wordCounts = rdd
      .flatMap(s -> Arrays.asList(s.toLowerCase().split(" ")))
      .mapToPair(word -> new Tuple2<>(word, 1))
      .reduceByKey((a, b) -> a + b);
```

之所以选用 Apache Spark 而不使用 MapReduce，原因之一是使用 Apache Spark 实现相同任务时，所需要的代码更少。比如，本步中用来从文本文档提取单词与词频的那些代码。对于这个任务，若使用 MapReduce 实现，则需要编写超过 100 行的代码。

10. 借助 wordCounts RDD，得到文档中的单词及其出现频数，并把它们放入 Map 中。然后，遍历 Map，把单词-频数成对打印出来。

```
    Map<String, Integer> wordMap = wordCounts.collectAsMap();
    for (Entry<String, Integer> entry : wordMap.entrySet()) {
        System.out.println("Word = " + entry.getKey() + ", Frequency
            = " + entry.getValue());
    }
```

11. 关闭前面创建的 sparkContext。

```
    sparkContext.close();
```

12. 关闭 main 方法与类。

```
    }
    }
```

示例的完整代码整理如下：

```
package com.data.big.mlib;

import java.util.Arrays;
import java.util.Map;
import java.util.Map.Entry;
import org.apache.spark.SparkConf;
import org.apache.spark.api.java.JavaPairRDD;
```

```java
import org.apache.spark.api.java.JavaRDD;
import org.apache.spark.api.java.JavaSparkContext;
import org.apache.spark.api.java.function.Function;
import scala.Tuple2;
public class SparkTest {
    public static void main( String[] args ){
        String inputFile = "data/shakespeare.txt";
        SparkConf configuration = new
           SparkConf().setMaster("local[4]").setAppName("My App");
        JavaSparkContext sparkContext = new
           JavaSparkContext(configuration);
        JavaRDD<String> rdd = sparkContext.textFile(inputFile).cache();
        long emptyLines = rdd.filter(new Function<String,Boolean>(){
            private static final long serialVersionUID = 1L;
            public Boolean call(String s){
                return s.length() == 0;
            }
        }).count();
        System.out.println("Empty Lines: " + emptyLines);
        JavaPairRDD<String, Integer> wordCounts = rdd
            .flatMap(s -> Arrays.asList(s.toLowerCase().split(" ")))
            .mapToPair(word -> new Tuple2<>(word, 1))
            .reduceByKey((a, b) -> a + b);
        Map<String, Integer> wordMap = wordCounts.collectAsMap();
        for (Entry<String, Integer> entry : wordMap.entrySet()) {
            System.out.println("Word = " + entry.getKey() + ", Frequency
                = " + entry.getValue());
        }
        sparkContext.close();
    }
}
```

运行上面代码，部分输出显示如下：

Empty Lines: 35941
....
....
Word = augustus, Frequency = 4
Word = bucklers, Frequency = 3
Word = guilty, Frequency = 66
Word = thunder'st, Frequency = 1

```
Word = hermia's, Frequency = 7
Word = sink, Frequency = 37
Word = burn, Frequency = 76
Word = relapse, Frequency = 2
Word = boar, Frequency = 16
Word = cop'd, Frequency = 2
..............................................................................
............................
```

> 在如下页面中刊载了一篇不错的文章，它鼓励用户使用 Apache Spark 来代替 MapReduce，感兴趣的读者，可以进一步了解。

7.5 使用 MLib 的 K 均值算法做聚类

本节中，我们将演示如何使用 MLib 的 K 均值算法对无标签数据点进行聚类。正如本章简介中所提到的，MLib 是 Apache Spark 的机器学习组件，它是 Apache Mahout 强有力的竞争对手，甚至优于 Apache Mahout。

准备工作

1. 我们将继续使用上一节（使用 Apache Spark 解决简单文本挖掘问题）创建的 Maven 项目。如果你还没有创建这个项目，请先参考上一节"准备工作"中步骤 1~6 创建该项目。

2. 前往 https://github.com/apache/spark/blob/master/data/mllib/kmeans_data.txt，下载数据，并将其保存在上一步所建项目的 data 文件夹之下，重命名为 km-data.txt。或者，先在项目的 data 文件夹中创建一个名为 km-data.txt 的文本文件，然后把上面 URL 中的数据粘贴到其中。

3. 在你创建的包中，创建一个名为 KMeansClusteringMlib.java 的 Java 类文件，双击打开它，准备编写代码。

   接下来，就要开始编写代码了。

操作步骤

1. 创建一个名为 KMeansClusteringMlib 的类。

```
public class KMeansClusteringMlib {
```

2. 在类中,开始编写 main 方法。

   ```
   public static void main( String[] args ){
   ```

3. 创建一个 Spark 配置,并使用该配置,创建一个 Spark context。请注意,如果我们使用了 `local[2]`,将启用最小并行设置。下面代码中,我们使用了 `local[4]`,允许应用程序运行 4 个线程。

   ```
   SparkConf configuration = new
    SparkConf().setMaster("local[4]").setAppName("K-means
      Clustering");
   JavaSparkContext sparkContext = new
     JavaSparkContext(configuration);
   ```

4. 接下来,开始加载与分析输入数据。

   ```
   String path = "data/km-data.txt";
   ```

5. JavaRDD 是一个分布式的对象集。创建一个 RDD 对象,以便读取数据文件。

   ```
   JavaRDD<String> data = sparkContext.textFile(path);
   ```

6. 然后,需要从前面的 RDD 读取数据值,这些值之间以空格键作为分隔。对这些数据值进行解析,并把它们读到另一个 RDD 中。

   ```
   JavaRDD<Vector> parsedData = data.map(
      new Function<String, Vector>() {
         private static final long serialVersionUID = 1L;

         public Vector call(String s) {
            String[] sarray = s.split(" ");
            double[] values = new double[sarray.length];
            for (int i = 0; i < sarray.length; i++)
               values[i] = Double.parseDouble(sarray[i]);
            return Vectors.dense(values);
         }
      }
   );
   parsedData.cache();
   ```

7. 接下来，为 K 均值聚类算法定义几个参数。这里，我们只使用两个簇集来分离数据点，最多 10 次迭代。使用解析的数据与参数值创建聚类器。

    ```
    int numClusters = 2;
    int iterations = 10;
    KMeansModel clusters = KMeans.train(parsedData.rdd(),
        numClusters, iterations);
    ```

8. 在聚类器集合中，计算误差平方和。

    ```
    double sse = clusters.computeCost(parsedData.rdd());
      System.out.println("Sum of Squared Errors within set = " +
        sse);
    ```

9. 最后，关闭 sparkContext、main 方法与类。

    ```
      sparkContext.close();
     }
    }
    ```

示例的完整代码整理如下：

```
package com.data.big.mlib;

import org.apache.spark.api.java.*;
import org.apache.spark.api.java.function.Function;
import org.apache.spark.mllib.clustering.KMeans;
import org.apache.spark.mllib.clustering.KMeansModel;
import org.apache.spark.mllib.linalg.Vector;
import org.apache.spark.mllib.linalg.Vectors;
import org.apache.spark.SparkConf;

public class KMeansClusteringMlib {
  public static void main( String[] args ){
    SparkConf configuration = new
      SparkConf().setMaster("local[4]").setAppName("K-means
        Clustering");
    JavaSparkContext sparkContext = new
      JavaSparkContext(configuration);
    //加载与解析数据
    String path = "data/km-data.txt";
    JavaRDD<String> data = sparkContext.textFile(path);
```

```java
        JavaRDD<Vector> parsedData = data.map(
            new Function<String, Vector>() {
                private static final long serialVersionUID = 1L;

                public Vector call(String s) {
                    String[] sarray = s.split(" ");
                    double[] values = new double[sarray.length];
                    for (int i = 0; i < sarray.length; i++)
                        values[i] = Double.parseDouble(sarray[i]);
                    return Vectors.dense(values);
                }
            }
        );
        parsedData.cache();

        //使用 K 均值把数据聚集到两个类中
        int numClusters = 2;
        int iterations = 10;
        KMeansModel clusters = KMeans.train(parsedData.rdd(),
            numClusters, iterations);

        //通过计算集合的误差平方和评估聚类
          Errors
        double sse = clusters.computeCost(parsedData.rdd());
        System.out.println("Sum of Squared Errors within set = " + sse);
        sparkContext.close();
    }
}
```

运行上面代码,得到如下输出:

```
Sum of Squared Errors within set = 0.11999999999994547
```

7.6 使用 MLib 创建线性回归模型

本节,我们将学习如何使用 MLib 来创建线性回归模型。

准备工作

1. 我们将继续使用 7.4 节(使用 Apache Spark 解决简单文本挖掘问题)中创建的 Maven 项目。如果你还没有创建这个项目,请先参考 7.6 节"准备工作"中步骤

1～6 创建该项目。

2. 前往 `https://github.com/apache/spark/blob/master/data/mllib/ridge-data/lpsa.data`，下载数据，并将其保存在上一步所建项目的 data 文件夹之下，重命名为 `lr-data.txt`。或者，先在项目的 data 文件夹中创建一个名为 `lr-data.txt` 的文本文件，然后把上面 URL 中的数据粘贴到其中。

3. 在你创建的包中，创建一个名为 `LinearRegressionMlib.java` 的 Java 类文件，双击打开它，准备编写代码。

接下来，就要开始编写代码了。

操作步骤

1. 创建一个名为 `LinearRegressionMlib` 的类。

   ```
   public class LinearRegressionMlib {
   ```

2. 在类中，开始编写 main 方法。

   ```
   public static void main(String[] args) {
   ```

3. 创建一个 Spark 配置，并使用该配置，创建一个 Spark context。请注意，如果我们使用 `local[2]`，这会启用最小并行设置。下面代码中，我们使用了 `local[4]`，允许应用程序运行 4 个线程。

   ```
   SparkConf configuration = new
     SparkConf().setMaster("local[4]").setAppName("Linear
       Regression");
   JavaSparkContext sparkContext = new
     JavaSparkContext(configuration);
   ```

4. 接下来，开始加载与分析输入数据。

   ```
   String inputData = "data/lr-data.txt";
   ```

5. JavaRDD 是一个分布式的对象集。创建一个 RDD 对象，以便读取数据文件。

   ```
   JavaRDD<String> data = sparkContext.textFile(inputData);
   ```

6. 然后，我们需要从上面的 RDD 读取数据值，输入的数据中包含两个部分或区段，它们之间以逗号作为分隔。在第二部分中，这些特征之间以空格键作为分隔。输入数据每一行的第一部分是标记点。解析这些数据值，并把它们读到另一个 RDD。使用这些特征，创建一个特征向量，并把它与标记点放在一起。

```java
JavaRDD<LabeledPoint> parsedData = data.map(
    new Function<String, LabeledPoint>() {
        private static final long serialVersionUID = 1L;

        public LabeledPoint call(String line) {
            String[] parts = line.split(",");
            String[] features = parts[1].split(" ");
            double[] featureVector = new
                double[features.length];
            for (int i = 0; i < features.length - 1; i++){
                featureVector[i] =
                    Double.parseDouble(features[i]);
            }
            return new LabeledPoint(Double.parseDouble(parts[0]),
                Vectors.dense(featureVector));
        }
    }
);
parsedData.cache();
```

7. 接下来，使用 10 次迭代创建线性回归模型，即使用特征向量、标签点、迭代数信息创建线性回归模型。

```java
int iterations = 10;
final LinearRegressionModel model =
    LinearRegressionWithSGD.train(JavaRDD.toRDD(parsedData),
        iterations);
```

8. 然后，使用创建好的模型进行预测，并把预测结果放入另外一个名为 predictions 的 RDD 变量中。给出一组特征，模型会产生一个预测值，返回预测结果与实际标签。请注意，此时我们得到的预测结果是针对训练集（lr-data.txt）中数据点所进行的预测。Tuple2 中既包含着回归模型的预测值，也包含着实际值。

```java
JavaRDD<Tuple2<Double, Double>> predictions = parsedData.map(
    new Function<LabeledPoint, Tuple2<Double, Double>>() {
        private static final long serialVersionUID = 1L;
```

```java
            public Tuple2<Double, Double> call(LabeledPoint point)
        {
            double prediction = model.predict(point.features());
            return new Tuple2<Double, Double>(prediction,
              point.label());
        }
    }
);
```

9. 最后，计算线性回归模型在训练数据上的均方误差。对于每个数据点，误差是指模型预测值与数据集中的实际值之差的平方。最后，对每个数据点的误差进行平均。

```java
double mse = new JavaDoubleRDD(predictions.map(
    new Function<Tuple2<Double, Double>, Object>() {
        private static final long serialVersionUID = 1L;

        public Object call(Tuple2<Double, Double> pair) {
            return Math.pow(pair._1() - pair._2(), 2.0);
        }
    }
).rdd()).mean();
System.out.println("training Mean Squared Error = " + mse);
```

10. 最后，关闭 sparkContext、main 方法与类。

```java
sparkContext.close();
    }
}
```

示例的完整代码整理如下：

```java
package com.data.big.mlib;

import scala.Tuple2;
import org.apache.spark.api.java.*;
import org.apache.spark.api.java.function.Function;
import org.apache.spark.mllib.linalg.Vectors;
import org.apache.spark.mllib.regression.LabeledPoint;
import org.apache.spark.mllib.regression.LinearRegressionModel;
import org.apache.spark.mllib.regression.LinearRegressionWithSGD;
import org.apache.spark.SparkConf;
```

```java
public class LinearRegressionMlib {

    public static void main(String[] args) {
        SparkConf configuration = new
            SparkConf().setMaster("local[4]").setAppName("Linear
                Regression");
        JavaSparkContext sparkContext = new
            JavaSparkContext(configuration);
        //加载与解析数据
        String inputData = "data/lr-data.txt";
        JavaRDD<String> data = sparkContext.textFile(inputData);
        JavaRDD<LabeledPoint> parsedData = data.map(
            new Function<String, LabeledPoint>() {
                private static final long serialVersionUID = 1L;

                public LabeledPoint call(String line) {
                    String[] parts = line.split(",");
                    String[] features = parts[1].split(" ");
                    double[] featureVector = new
                      double[features.length];
                    for (int i = 0; i < features.length - 1; i++){
                        featureVector[i] =
                            Double.parseDouble(features[i]);
                    }
                    return new LabeledPoint(Double.parseDouble(parts[0]),
                        Vectors.dense(featureVector));
                }
            }
        );
        parsedData.cache();

        //创建模型
        int iterations = 10;
        final LinearRegressionModel model =
            LinearRegressionWithSGD.train(JavaRDD.toRDD(parsedData),
                iterations);

        //评估模型在训练样本的表现并计算训练误差
            error
        JavaRDD<Tuple2<Double, Double>> predictions = parsedData.map(
            new Function<LabeledPoint, Tuple2<Double, Double>>() {
                private static final long serialVersionUID = 1L;
```

```java
            public Tuple2<Double, Double> call(LabeledPoint point) {
                double prediction = model.predict(point.features());
                return new Tuple2<Double, Double>(prediction,
                  point.label());
            }
        }
    );
    double mse = new JavaDoubleRDD(predictions.map(
        new Function<Tuple2<Double, Double>, Object>() {
            private static final long serialVersionUID = 1L;

            public Object call(Tuple2<Double, Double> pair) {
                return Math.pow(pair._1() - pair._2(), 2.0);
            }
        }
    ).rdd()).mean();
    System.out.println("training Mean Squared Error = " + mse);
    sparkContext.close();
    }
}
```

运行上面代码，得到如下输出结果：

training Mean Squared Error = 6.487093790021849

7.7 使用 MLib 的随机森林模型对数据点进行分类

本节，我们将演示使用 MLib 的随机森林算法对数据点进行分类的方法。

准备工作

1. 我们将继续使用 7.4 节（使用 Apache Spark 解决简单文本挖掘问题）中创建的 Maven 项目。如果你还没有创建这个项目，请先参考 7.6 节"准备工作"中步骤 1～6 创建该项目。

2. 前往 https://github.com/apache/spark/blob/master/data/mllib/ sample_binary_ classification_data.txt，下载数据，并将其保存在上一步所建项目的 data 文件夹之下，重命名为 rf-data.txt。或者，先在项目的 data 文件夹中创建一个名为 rf-data.txt 的文本文件，然后把上面 URL 中的数

据粘贴到其中。

3. 在你创建的包中，创建一个名为 RandomForestMlib.java 的 Java 类文件，双击打开它，准备编写代码。

操作步骤

1. 创建一个名为 RandomForestMlib 的类。

   ```
   public class RandomForestMlib {
   ```

2. 在类中，开始编写 main 方法。

   ```
   public static void main(String args[]){
   ```

3. 创建一个 Spark 配置，并使用该配置，创建一个 Spark context。请注意，如果我们使用 local[2]，这会启用最小并行设置。下面代码中，我们使用了 local[4]，表示允许应用程序运行 4 个线程。

   ```
   SparkConf configuration = new
     SparkConf().setMaster("local[4]").setAppName("Random
       Forest");
   JavaSparkContext sparkContext = new
     JavaSparkContext(configuration);
   ```

4. 接下来，开始加载与分析输入数据。

   ```
   String input = "data/rf-data.txt";
   ```

5. 通过把输入文件加装为 LibSVM 文件，读取数据，并将其放入一个 RDD 之中。

   ```
   JavaRDD<LabeledPoint> data =
     MLUtils.loadLibSVMFile(sparkContext.sc(),
       input).toJavaRDD();
   ```

6. 我们将使用 70%的数据来训练模型，把剩余的 30%用作模型的测试数据。并且数据的选择是随机的。

   ```
   JavaRDD<LabeledPoint>[] dataSplits = data.randomSplit(new
     double[]{0.7, 0.3});
   JavaRDD<LabeledPoint> trainingData = dataSplits[0];
   ```

```
JavaRDD<LabeledPoint> testData = dataSplits[1];
```

7. 接下来，我们将配置几个参数，建立随机森林，以便根据训练数据生成模型。首先，需要定义数据点所拥有的分类数。然后，需要为名义特征创建一个 Map。我们可以定义森林中树的棵数。如果你不知道该如何为分类器选择特征子集过程，你可以使用 auto。其余 4 个参数也是森林结构所必需的。

```
Integer classes = 2;
HashMap<Integer, Integer> nominalFeatures = new
 HashMap<Integer, nteger>();
Integer trees = 3;
String featureSubsetProcess = "auto";
String impurity = "gini";
Integer maxDepth = 3;
Integer maxBins = 20;
Integer seed = 12345;
```

8. 使用这些参数，创建一个 RandomForest 分类器。

```
final RandomForestModel rf =
  RandomForest.trainClassifier(trainingData, classes,
    nominalFeatures, trees, featureSubsetProcess, impurity,
      maxDepth, maxBins, seed);
```

9. 接下来，使用创建好的模型，依据给定的特征向量预测数据点的类标签。对于每个数据点，Tuple2<Double,Double>中包含着预测值与真实类值。

```
JavaPairRDD<Double, Double> label =
  testData.mapToPair(new PairFunction<LabeledPoint, Double,
    Double>() {
      private static final long serialVersionUID = 1L;

      public Tuple2<Double, Double> call(LabeledPoint p) {
        return new Tuple2<Double, Double>
          (rf.predict(p.features()), p.label());
      }
});
```

10. 最后，计算预测误差。我们可以简单地统计预测值与真实值不匹配的次数，然后除以测试实例总数，得到平均值。

```
        Double error =
          1.0 * label.filter(new Function<Tuple2<Double, Double>,
            Boolean>() {
                private static final long serialVersionUID = 1L;

                public Boolean call(Tuple2<Double, Double> pl) {
                    return !pl._1().equals(pl._2());
                }
        }).count() / testData.count();
```

11. 把测试误差在控制台中打印出来。或许，你还想看一看从训练数据中学到的 RandomForest 模型，使用如下语句，即可将其打印出来。

```
        System.out.println("Test Error: " + error);
        System.out.println("Learned classification forest model:\n" +
          rf.toDebugString());
```

12. 关闭 sparkContext、main 方法与类。

```
        sparkContext.close();
    }
}
```

示例的完整代码整理如下：

```
package com.data.big.mlib;

import scala.Tuple2;
import java.util.HashMap;
import org.apache.spark.SparkConf;
import org.apache.spark.api.java.JavaPairRDD;
import org.apache.spark.api.java.JavaRDD;
import org.apache.spark.api.java.JavaSparkContext;
import org.apache.spark.api.java.function.Function;
import org.apache.spark.api.java.function.PairFunction;
import org.apache.spark.mllib.regression.LabeledPoint;
import org.apache.spark.mllib.tree.RandomForest;
import org.apache.spark.mllib.tree.model.RandomForestModel;
import org.apache.spark.mllib.util.MLUtils;

public class RandomForestMlib {
   public static void main(String args[]){
```

```java
SparkConf configuration = new
    SparkConf().setMaster("local[4]").setAppName("Random Forest");
JavaSparkContext sparkContext = new
    JavaSparkContext(configuration);

//加载与分析数据文件
String input = "data/rf-data.txt";
JavaRDD<LabeledPoint> data =
    MLUtils.loadLibSVMFile(sparkContext.sc(), input).toJavaRDD();
//把数据划分为训练集与测试集(抽取30%作为测试)
JavaRDD<LabeledPoint>[] dataSplits = data.randomSplit(new
    double[]{0.7, 0.3});
JavaRDD<LabeledPoint> trainingData = dataSplits[0];
JavaRDD<LabeledPoint> testData = dataSplits[1];

//训练随机森林模型
Integer classes = 2;
HashMap<Integer, Integer> nominalFeatures = new HashMap<Integer,
    Integer>();//空 categoricalFeaturesInfo 表示所有特征都是连续的
        features are continuous.
Integer trees = 3; //实际设置得更多
String featureSubsetProcess = "auto"; //让算法自己选择
String impurity = "gini";
Integer maxDepth = 3;
Integer maxBins = 20;
Integer seed = 12345;
final RandomForestModel rf =
    RandomForest.trainClassifier(trainingData, classes,
        nominalFeatures, trees, featureSubsetProcess, impurity,
        maxDepth, maxBins, seed);
//在测试实例上评估模型,并计算测试误差
JavaPairRDD<Double, Double> label =
    testData.mapToPair(new PairFunction<LabeledPoint, Double,
        Double>() {
        private static final long serialVersionUID = 1L;

        public Tuple2<Double, Double> call(LabeledPoint p) {
            return new Tuple2<Double, Double>
                (rf.predict(p.features()), p.label());
        }
```

```
                });
            Double error =
                1.0 * label.filter(new Function<Tuple2<Double, Double>,
                    Boolean>() {
                        private static final long serialVersionUID = 1L;

                        public Boolean call(Tuple2<Double, Double> pl) {
                            return !pl._1().equals(pl._2());
                        }
                }).count() / testData.count();
            System.out.println("Test Error: " + error);
            System.out.println("Learned classification forest model:\n" +
                rf.toDebugString());
            sparkContext.close();
        }
    }
```

运行上面的代码，得到如下输出：

```
Test Error: 0.034482758620689655
Learned classification forest model:
TreeEnsembleModel classifier with 3 trees

  Tree 0:
    If (feature 427 <= 0.0)
     If (feature 407 <= 0.0)
      Predict: 0.0
     Else (feature 407 > 0.0)
      Predict: 1.0
    Else (feature 427 > 0.0)
     Predict: 0.0
  Tree 1:
    If (feature 405 <= 0.0)
     If (feature 624 <= 253.0)
      Predict: 0.0
     Else (feature 624 > 253.0)
      If (feature 650 <= 0.0)
       Predict: 0.0
      Else (feature 650 > 0.0)
       Predict: 1.0
    Else (feature 405 > 0.0)
     If (feature 435 <= 0.0)
      If (feature 541 <= 0.0)
```

```
        Predict: 1.0
      Else (feature 541 > 0.0)
        Predict: 0.0
    Else (feature 435 > 0.0)
      Predict: 1.0
  Tree 2:
    If (feature 271 <= 72.0)
     If (feature 323 <= 0.0)
      Predict: 0.0
     Else (feature 323 > 0.0)
      Predict: 1.0
    Else (feature 271 > 72.0)
     If (feature 414 <= 0.0)
      If (feature 159 <= 124.0)
       Predict: 0.0
      Else (feature 159 > 124.0)
       Predict: 1.0
     Else (feature 414 > 0.0)
       Predict: 0.0
```

第 8 章 数据深度学习

本章包含如下内容：

- 使用 DL4j 创建 Word2vec 神经网络；
- 使用 DL4j 创建深度信念神经网络；
- 使用 DL4j 创建深度自动编码器。

8.1 简介

简单地说，深度学习是指带有多个层的神经网络，也叫深度神经网络学习或非监督特征学习。我相信深度学习会成为机器学习实践者与数据科学家又一个得力的"助手"，这是因为它在解决实际的数据问题方面表现出强大的能力。

DL4j（Deep Learning for Java）是一个面向 JVM 开源的、分布式 Java 深度学习库。伴随着 DL4j 一起而来的还有其他许多库，如下所示。

- Deeplearning4J：神经网络平台。
- ND4J：面向 JVM 的 NumPy。
- DataVec：进行机器学习 ETL 处理的工具。
- JavaCPP：Java 与本地 C++的桥梁。
- Arbiter：机器学习算法评估工具。
- RL4J：面向 JVM 的深度增强学习。

然而，本书中，我们只重点讲解 DL4j 的几个关键方面。具体地说，我们将讲解

Word2vec 算法、real-world NLP、信息提取问题、深度信念神经网络、深度自动编码器及它们的用法。感兴趣的读者，建议你访问 https://github.com/deeplearning4j/dl4j-examples，获取更多例子。请注意，本章中的所有代码都是基于这些示例的，并且在 GitHub 上可以找到。

另外，还请你注意，本章使用了很大篇幅来演示如何安装 DL4j 库，这是因为安装 DL4j 的过程非常复杂。并且，大家要足够细心，才能保证成功运行本书以及你们自己编写的代码。

在学习本章内容之前，需要先做如下两项准备：下载 Java Developer version 1.7 或更高版本（我使用的是 1.8 版本）与 Apache Maven。本章中的所有代码都是使用 Eclipse Java IDE（我使用的是 Eclipse Mars）编写的。虽然 https://deeplearning4j.org/quickstart 包含大量安装 DL4j 的参考资料，但大部分讲的都是另一个 IDE，即 IntelliJ。

开始学习之前，先做如下准备。

1. 在使用 DL4j 之前，需要先安装 Apache Maven，它是一种软件项目管理工具。写作本书之时，Apache Maven 最新版本为 3.3.9，建议各位使用这个版本。

2. 前往 Apache Maven 官网，把二进制 zip 文件下载到本地系统中（见图 8-1）。

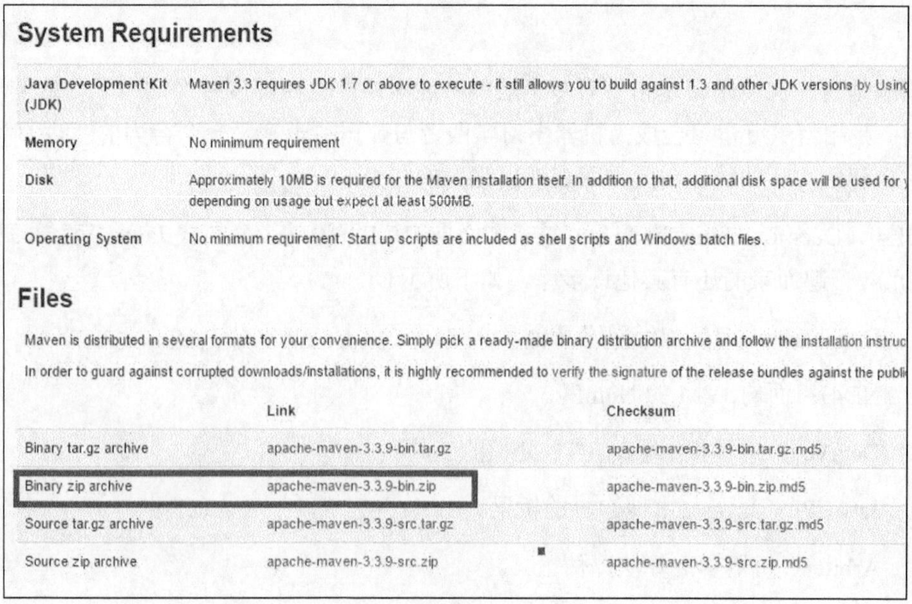

图 8-1

3. 下载完成后，进行解压缩，得到如图 8-2 所示的目录结构。

bin	2017-02-06 1:51 PM	File folder	
boot	2017-02-06 1:51 PM	File folder	
conf	2017-02-06 1:51 PM	File folder	
lib	2017-02-06 1:51 PM	File folder	
LICENSE	2015-11-10 11:44 ...	File	19 KB
NOTICE	2015-11-10 11:44 ...	File	1 KB
README.txt	2015-11-10 11:38 ...	Text Document	3 KB

图 8-2

4. 接下来，需要把 bin 文件夹的路径放到类路径中。为此，右键单击"我的电脑"图标，在弹出菜单中，选择"属性"，再选择"高级系统属性"，然后在"系统属性"中，选择"环境变量"，如图 8-3 所示。

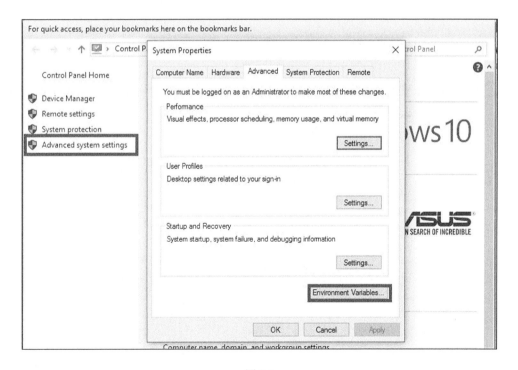

图 8-3

5. 在"环境变量"对话框的"系统变量"中，选择 Path 变量，单击"编辑"按钮（见图 8-4）。

232 第 8 章 数据深度学习

图 8-4

6. 在"编辑环境变量"窗口中,单击"编辑"按钮,如图 8-5 所示,把 Maven 的 bin 目录添加进去,单击"确定"按钮,使更改生效。

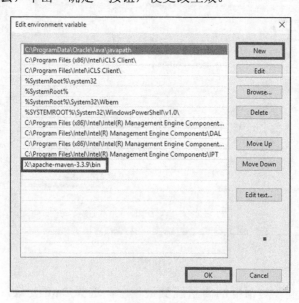

图 8-5

7. 返回到"环境变量"窗口，创建系统变量 JAVA_HOME。为此，需要在"系统变量"版块中，单击"新建"按钮（见图 8-6）。

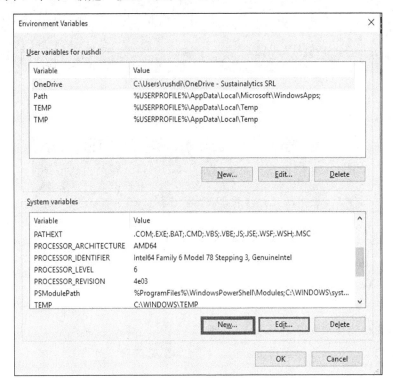

图 8-6

8. 在"新建系统变量"对话框中，在"变量名"中输入 JAVA_HOME，在变量值中输入 JDK 路径（请注意，不是 bin 文件夹路径）。

> 请注意，若想运行本章中的示例代码，你电脑中安装的 Java 版本应该不低于 7。

9. 如图 8-7 所示，单击"确定"按钮，完成设置。关闭所有打开的窗口。

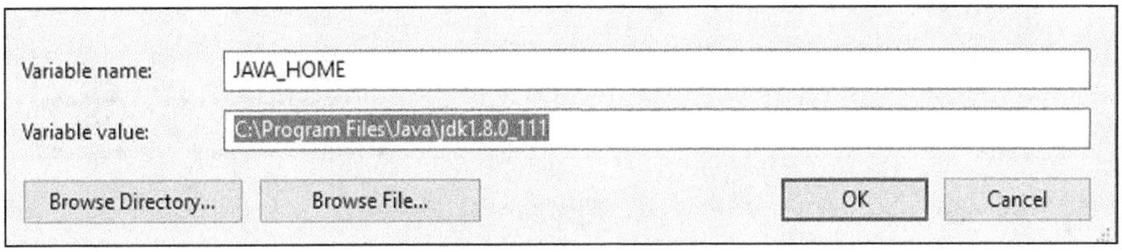

图 8-7

10. 接下来，使用如图 8-8 所示的 mvn -v 命令检查 Maven 安装是否正确。

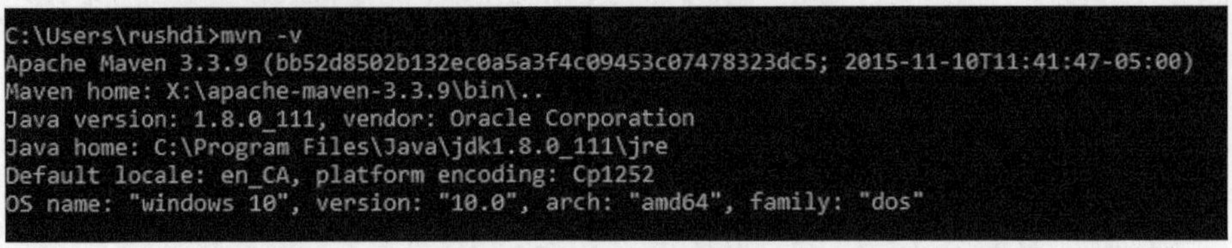

图 8-8

11. 然后，使用如图 8-9 所示的 java -version 命令，查看系统中安装的 Java 版本。

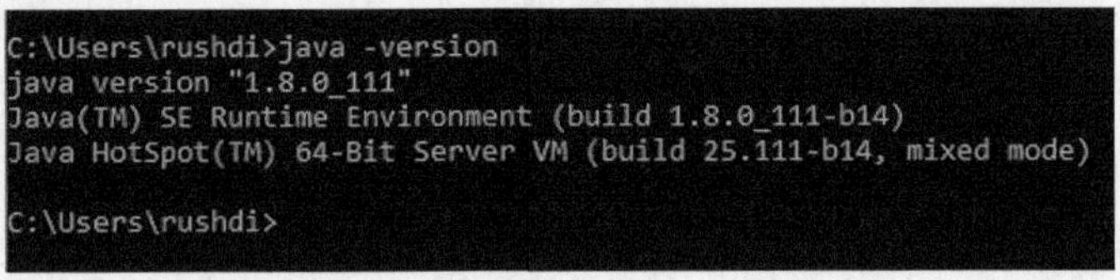

图 8-9

12. 打开 Eclipse IDE（我使用的是 Eclipse Mars），在菜单栏中，依次选择 "File|New|Other..." 菜单（见图 8-10）。

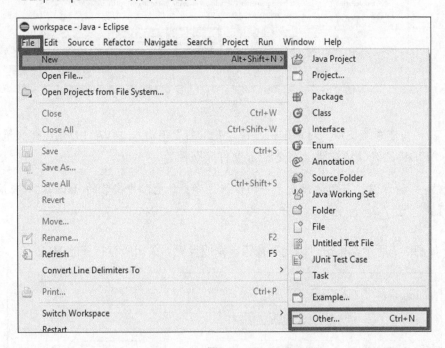

图 8-10

13. 如图 8-11 所示，在"新建"窗口中，展开 Maven 选项，选择 Maven 项目，单击

"**Next**"按钮。

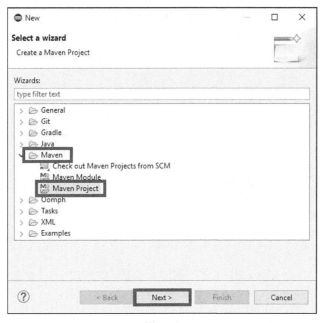

图 8-11

14. 不断单击"**Next**"按钮，直到出现如下窗口。在 **Group Id** 与 **Artifact Id** 中，填入相应名称，单击"**Finish**"按钮（见图 8-12）。

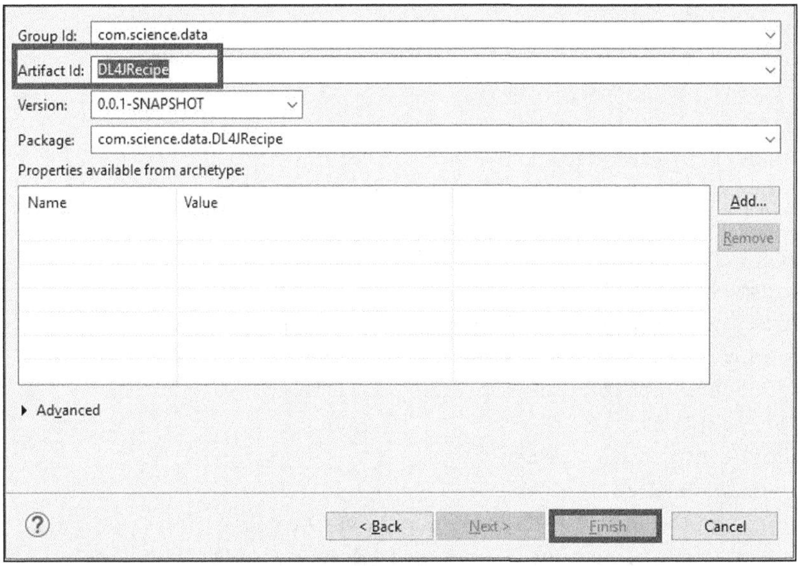

图 8-12

15. 经过上面一系列操作之后，创建好如图 8-13 所示的项目。双击项目名称，把项目展开，你将会看到一个名为 POM.xml 的 XML 文件。

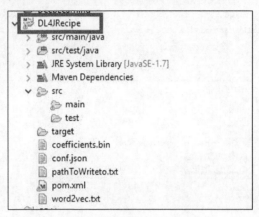

图 8-13

16. 双击 pom.xml 文件，打开它，删除其中所有的内容，并把如下内容粘贴进去。

```
<project xmlns="http://maven.apache.org/POM/4.0.0"
    xmlns:xsi="http://www.w3.org/2001/XMLSchema-instance"
 xsi:schemaLocation="http://maven.apache.org/POM/4.0.0
 http://maven.apache.org/xsd/maven-4.0.0.xsd">
<modelVersion>4.0.0</modelVersion>

<groupId>org.deeplearning4j</groupId>
<artifactId>deeplearning4j-examples</artifactId>
<version>0.4-rc0-SNAPSHOT</version>

<name>DeepLearning4j Examples</name>
<description>Examples of training different data
    sets</description>
<properties>
 <nd4j.version>0.4-rc3.7</nd4j.version>
 <dl4j.version> 0.4-rc3.7</dl4j.version>
 <canova.version>0.0.0.13</canova.version>
 <jackson.version>2.5.1</jackson.version>
</properties>
<distributionManagement>
    <snapshotRepository>
        <id>sonatype-nexus-snapshots</id>
        <name>Sonatype Nexus snapshot repository</name>
<url>https://oss.sonatype.org/content/repositories/snapshots</url>
```

```xml
        </snapshotRepository>
        <repository>
            <id>nexus-releases</id>
            <name>Nexus Release Repository</name>
        <url>http://oss.sonatype.org/service/local/
            staging/deploy/maven2/</url>
        </repository>
         </distributionManagement>
        <dependencyManagement>
         <dependencies>
            <dependency>
                <groupId>org.nd4j</groupId>
                <artifactId>nd4j-x86</artifactId>
                <version>${nd4j.version}</version>
            </dependency>
         </dependencies>
        </dependencyManagement>
    <dependencies>
    <dependency>
        <groupId>org.deeplearning4j</groupId>
        <artifactId>deeplearning4j-nlp</artifactId>
        <version>${dl4j.version}</version>
    </dependency>

    <dependency>
        <groupId>org.deeplearning4j</groupId>
        <artifactId>deeplearning4j-core</artifactId>
        <version>${dl4j.version}</version>
    </dependency>
    <dependency>
        <groupId>org.deeplearning4j</groupId>
        <artifactId>deeplearning4j-ui</artifactId>
        <version>${dl4j.version}</version>
    </dependency>
    <dependency>
        <groupId>org.nd4j</groupId>
        <artifactId>nd4j-x86</artifactId>
        <version>${nd4j.version}</version>
    </dependency>
    <dependency>
        <artifactId>canova-nd4j-image</artifactId>
        <groupId>org.nd4j</groupId>
```

```xml
            <version>${canova.version}</version>
        </dependency>
        <dependency>
            <artifactId>canova-nd4j-codec</artifactId>
            <groupId>org.nd4j</groupId>
            <version>${canova.version}</version>
        </dependency>
        <dependency>
            <groupId>com.fasterxml.jackson.dataformat</groupId>
            <artifactId>jackson-dataformat-yaml</artifactId>
            <version>${jackson.version}</version>
        </dependency>
    </dependencies>
    <build>
      <plugins>
        <plugin>
            <groupId>org.codehaus.mojo</groupId>
            <artifactId>exec-maven-plugin</artifactId>
            <version>1.4.0</version>
            <executions>
                <execution>
                    <goals>
                        <goal>exec</goal>
                    </goals>
                </execution>
            </executions>
            <configuration>
                <executable>java</executable>
            </configuration>
        </plugin>
        <plugin>
            <groupId>org.apache.maven.plugins</groupId>
            <artifactId>maven-shade-plugin</artifactId>
            <version>1.6</version>
            <configuration>
  <createDependencyReducedPom>true</createDependencyReducedPom>
        <filters>
        <filter>
        <artifact>*:*</artifact>
        <excludes>
            <exclude>org/datanucleus/**</exclude>
              <exclude>META-INF/*.SF</exclude>
```

```xml
                <exclude>META-INF/*.DSA</exclude>
                <exclude>META-INF/*.RSA</exclude>
          </excludes>
                </filter>
              </filters>
            </configuration>
            <executions>
                <execution>
                    <phase>package</phase>
                    <goals>
                        <goal>shade</goal>
                    </goals>
                    <configuration>
                        <transformers>
     <transformer implementation="org.apache.maven.plugins.
        shade.resource.AppendingTransformer">
     <resource>reference.conf</resource>
                        </transformer>
     <transformer implementation="org.apache.maven.plugins.
            shade.resource.ServicesResourceTransformer"/>
     <transformer implementation="org.apache.maven.plugins.
            shade.resource.ManifestResourceTransformer">
                        </transformer>
                        </transformers>
                    </configuration>
                </execution>
            </executions>
        </plugin>

        <plugin>
            <groupId>org.apache.maven.plugins</groupId>
            <artifactId>maven-compiler-plugin</artifactId>
            <configuration>
                <source>1.7</source>
                <target>1.7</target>
            </configuration>
        </plugin>
    </plugins>
  </build>
</project>
```

17. 这会下载所有必需的文件，如图 8-14 所示。接下来，准备编写一些代码。

图 8-14

18. 前往 `https://github.com/deeplearning4j/dl4j-examples/tree/master/dl4j-examples/src/main/resources`，把 `raw_sentences.txt` 文件下载到 C:/ 之下（见图 8-15）。

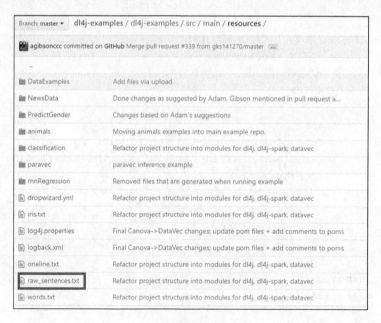

图 8-15

8.2 使用 DL4j 创建 Word2vec 神经网络

我们可以把 Word2vec 看成一个两层的神经网络，用来处理自然文本，把单词转换成向量形式。它的典型用法是算法接收文本语料库输入，然后输出语料库中单词所对应的一系列特征向量。严格来说，Word2vec 不是一个深度神经网络，它只是把文本转换成数字形式，以供深度神经网络进行读取与理解。本节，我们将学习如何使用 DL4j 这个流行的 Java 深度学习库对原始文本应用 Word2vec。

操作方法

1. 创建一个名为 Word2VecRawTextExample 的类。

    ```
    public class Word2VecRawTextExample {
    ```

2. 为这个类创建一个 Logger。在使用 Maven 创建项目时，Logger 已经包含在你的项目之中了。

    ```
    private static Logger log =
      LoggerFactory.getLogger(Word2VecRawTextExample.class);
    ```

3. 开始编写 main 方法。

    ```
    <dependency><groupId>org.nd4j</groupId><artifactId>nd4jnative</artifactId><version>0.7.2</version></dependency>
    ```

    ```
    public static void main(String[] args) throws Exception {
    ```

4. 前面我们已经下载了 raw_sentences.txt 这个文本文件，要做的第一件事是获取这个文件的路径。

    ```
    String filePath = "c:/raw_sentences.txt";
    ```

5. 接下来，获取.txt 文件中的原始句子，使用迭代器对它们进行遍历，并且进行预处理（比如转换成小写，去除每一行前后的空格）。

    ```
    log.info("Load & Vectorize Sentences....");
    SentenceIterator iter =
      UimaSentenceIterator.createWithPath(filePath);
    ```

6. Word2vec 使用单词或标记,不使用句子。因此,接下来的任务是对原始文本进行切分。

   ```
   TokenizerFactory t = new DefaultTokenizerFactory();
   t.setTokenPreProcessor(new CommonPreprocessor());
   ```

7. DL4j 提供了词汇缓存机制,用来处理一般的自然语言任务,比如 TF-IDF。InMemoryLookupCache 是一个参考实现。

   ```
   InMemoryLookupCache cache = new InMemoryLookupCache();
     WeightLookupTable table = new InMemoryLookupTable.Builder()
           .vectorLength(100)
           .useAdaGrad(false)
           .cache(cache)
           .lr(0.025f).build();
   ```

8. 至此,数据就准备好了。接下来,我们要配置 Word2vec 神经网络。

   ```
   log.info("Building model....");
     Word2Vec vec = new Word2Vec.Builder()
           .minWordFrequency(5).iterations(1)
           .layerSize(100).lookupTable(table)
           .stopWords(new ArrayList<String>())
           .vocabCache(cache).seed(42)
           .windowSize(5).iterate(iter).tokenizerFactory(t).build();
   ```

   minWordFrequency 是单词在语料库中最少出现的次数。本节中,如果单词出现次数少于 5 次,就不会学习它。单词必须出现在多个上下文环境中,以便学习与它们有关的有用特征。如果语料库非常大,提高单词最少出现次数是合理的做法。LayerSize 表示单词向量中的特征数,或者特征空间中的维数。接下来,启动神经网络训练拟合模型。

   ```
   log.info("Fitting Word2Vec model....");
   vec.fit();
   ```

9. 把神经网络产生的单词向量写入一个输出文件中。本例中,输出文件是 c:/word2vec.txt。

   ```
   <dependency><groupId>org.nd4j</groupId><artifactId>nd4jnative</artifactId><version>0.7.2</version></dependency>
   log.info("Writing word vectors to text file....");
   WordVectorSerializer.writeWordVectors(vec, "c:/word2vec.txt");
   ```

10. 我们还可以对特征向量的质量进行评估。vec.wordsNearest("word1",

numWordsNearest）语句会把神经网络判断为语义类似的单词聚集在一起，并提供给我们。并且我们可以通过 wordsNearest()函数的第二个参数设置想要的最近单词数。vec.similarity("word1","word2")用来返回所输入的两个单词的余弦相似度，其值越接近于 1，这两个单词就越类似。

```
log.info("Closest Words:");
Collection<String> lst = vec.wordsNearest("man", 5);
System.out.println(lst);
double cosSim = vec.similarity("cruise", "voyage");
System.out.println(cosSim);
```

11. 上面这几行输出如下：

```
[family, part, house, program, business]
1.0000001192092896
```

12. 关闭 main 方法与类。

```
    }
}
```

工作原理

1. 在 Eclipse 中的项目名上单击鼠标右键，选择 **New**，然后选择 **Package**。如图 8-16 所示，输入包名为 word2vec.chap8.science.data，单击 **Finish** 按钮。

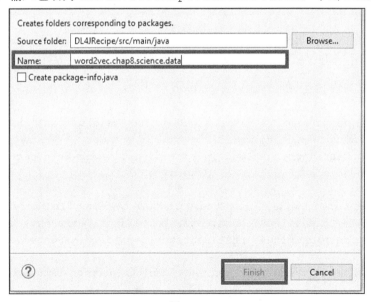

图 8-16

2. 创建好包之后，右键单击包名，依次选择 **New|Class**，输入类名为 Word2VecRawText Example，单击"**Finish**"按钮（见图 8-17）。

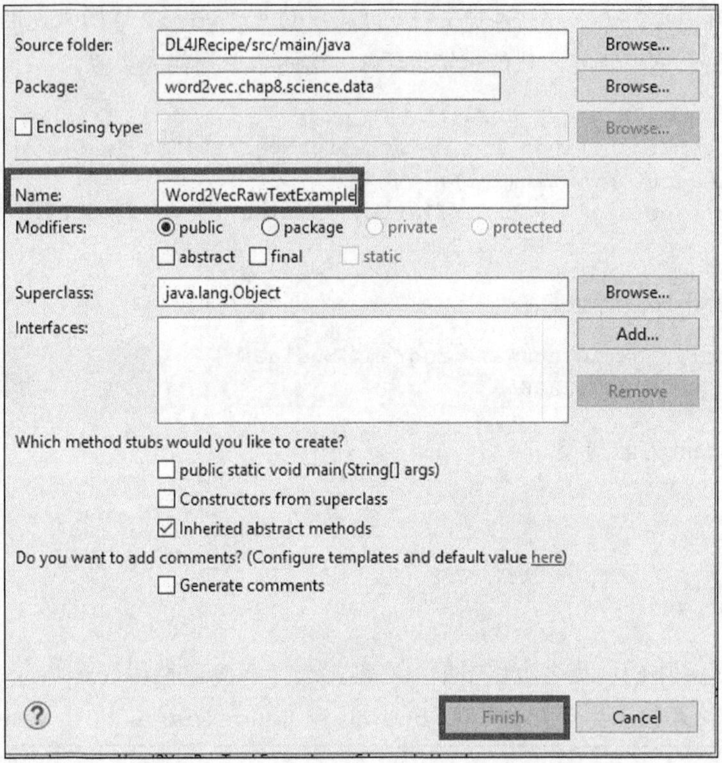

图 8-17

在编辑器中，复制与粘贴如下代码。

```
package word2vec.chap8.science.data;

import org.deeplearning4j.models.embeddings.WeightLookupTable;
import org.deeplearning4j.models.embeddings.inmemory.InMemoryLookupTable;
import org.deeplearning4j.models.embeddings.loader.WordVectorSerializer;
import org.deeplearning4j.models.word2vec.Word2Vec;
import org.deeplearning4j.models.word2vec.wordstore.inmemory.InMemoryLookupCache;
import org.deeplearning4j.text.sentenceiterator.SentenceIterator;
import org.deeplearning4j.text.sentenceiterator.UimaSentenceIterator;
import org.deeplearning4j.text.tokenization.tokenizer.preprocessor.CommonPreprocessor;
import
```

```
org.deeplearning4j.text.tokenization.tokenizerfactory.DefaultTokenizerFacto
ry;
import
org.deeplearning4j.text.tokenization.tokenizerfactory.TokenizerFactory;

import org.slf4j.Logger;
import org.slf4j.LoggerFactory;

import java.util.ArrayList;
import java.util.Collection;

public class Word2VecRawTextExample {

    private static Logger log =
LoggerFactory.getLogger(Word2VecRawTextExample.class);

    public static void main(String[] args) throws Exception {
        //获取文本文件路径
        String filePath = "c:/raw_sentences.txt";

        log.info("Load & Vectorize Sentences....");
        //去除每行前后的空格
        SentenceIterator iter =
            UimaSentenceIterator.createWithPath(filePath);
        //根据数据行中的空格进行切分，获取单词
        TokenizerFactory t = new DefaultTokenizerFactory();
        t.setTokenPreProcessor(new CommonPreprocessor());

        InMemoryLookupCache cache = new InMemoryLookupCache();
        WeightLookupTable table = new InMemoryLookupTable.Builder()
                .vectorLength(100)
                .useAdaGrad(false)
                .cache(cache)
                .lr(0.025f).build();

        log.info("Building model....");
        Word2Vec vec = new Word2Vec.Builder()
            .minWordFrequency(5).iterations(1)
            .layerSize(100).lookupTable(table)
            .stopWords(new ArrayList<String>())
            .vocabCache(cache).seed(42)
            .windowSize(5).iterate(iter).tokenizerFactory(t).build();
```

```
        log.info("Fitting Word2Vec model....");
        vec.fit();

        log.info("Writing word vectors to text file....");
        //写单词
        WordVectorSerializer.writeWordVectors(vec, "word2vec.txt");

        log.info("Closest Words:");
        Collection<String> lst = vec.wordsNearest("man", 5);
        System.out.println(lst);
        double cosSim = vec.similarity("cruise", "voyage");
        System.out.println(cosSim);
    }
}
```

更多内容

- `minWordFrequency`：指定某个单词在语料库中最少出现次数。本节中，如果单词出现次数少于 5 次，就不会学习它。单词必须出现在多个上下文环境中，以便学习与它们有关的有用特征。如果语料库非常大，提高单词最少出现次数是合理的做法。

- `iterations`：指定神经网络为一批数据更新自身系数的次数。设置得太小，可能会导致学习不充分，而设置太大又会导致网络训练时间太长。

- `LayerSize`：表示单词向量中的特征数，或者特征空间中的维数。

- `interate`：方法告知网络当前正在哪批数据上进行训练。

- `tokenizer`：把当前批中的单词传递给它。

8.3　使用 DL4j 创建深度信念神经网络

我们可以把深度信念网络定义为限制玻尔兹曼机的堆叠，每个 RBM 层都与前一个和后一个层进行通信。本节中，我们将学习如何创建这种网络。为了简便起见，我们的讨论只限于有 4 个神经网络与一个隐藏层的情形。因此，严格来说，本节中我们创建的网络不是深度信念神经网络，鼓励大家添加更多的隐藏层。

操作步骤

1. 创建一个名为 `DBNIrisExample` 的类。

```
public class DBNIrisExample {
```

2. 为这个类创建一个 Logger，用来记录信息。

```
private static Logger log =
  LoggerFactory.getLogger(DBNIrisExample.class);
```

3. 开始编写 main 方法。

```
public static void main(String[] args) throws Exception {
```

4. 首先，为 Nd4j 类设置两个参数，第一个参数是要打印的最大片（slice）数，第二个是每片的最大元素个数。这里，把它们全部设置为-1。

```
Nd4j.MAX_SLICES_TO_PRINT = -1;
Nd4j.MAX_ELEMENTS_PER_SLICE = -1;
```

5. 接着，设置其他参数。

```
final int numRows = 4;
 final int numColumns = 1;
  int outputNum = 3;
  int numSamples = 150;
  int batchSize = 150;
  int iterations = 5;
  int splitTrainNum = (int) (batchSize * .8);
  int seed = 123;
  int listenerFreq = 1;
```

- 在 DL4j 中，输入数据可以是二维数据，因此我们需要指定数据的行数与列数。这里之所以把列数设置为 1 是因为 Iris 数据集是一维的。
- 上面的代码中，numSamples 是总数据数，batchSize 是每批的数据数。
- splitTrainNum 用来分配训练与测试数据。这里，我们把数据集的 80%用作训练数据，其余数据用作测试数据。
- listenerFreq 是查看注册到进程的损失函数值的频率。这里，我们将其设置为 1，表示在每个 epoch 后注册值。

6. 编写如下代码，使用指定的批大小与样本数信息自动加载 Iris 数据集。

```
log.info("Load data....");
```

```
DataSetIterator iter = new IrisDataSetIterator(batchSize,
    numSamples);
```

7. 对数据进行格式化。

   ```
   DataSet next = iter.next();
   next.normalizeZeroMeanZeroUnitVariance();
   ```

8. 接着，把数据划分为训练数据与测试数据。进行划分时，使用随机种子，并且把 ENFORCE_NUMERICAL_STABILITY 设置为真。

   ```
   log.info("Split data....");
   SplitTestAndTrain testAndTrain =
       next.splitTestAndTrain(splitTrainNum, new Random(seed));
   DataSet train = testAndTrain.getTrain();
   DataSet test = testAndTrain.getTest();
   Nd4j.ENFORCE_NUMERICAL_STABILITY = true;
   ```

9. 接下来，使用如下代码创建模型。

   ```
   MultiLayerConfiguration conf = new
       NeuralNetConfiguration.Builder()
       .seed(seed)
       .iterations(iterations)
       .learningRate(1e-6f)
       .optimizationAlgo(OptimizationAlgorithm.CONJUGATE_GRADIENT)
       .l1(1e-1).regularization(true).l2(2e-4)
       .useDropConnect(true)
       .list(2)
   ```

10. 让我们一起分析一下上述代码。

    - 先使用 seed() 方法，确定调整权重的初始值。
    - 然后，为预测或分类设置训练迭代数。
    - 定义优化步长，选择反向传播算法计算梯度。
    - 最后，在 list() 方法中，设置参数为 2，指定神经网络层数（包含输入层）。

11. 然后，添加如下方法到上一步的代码中。这部分代码用来创建神经网络的第一个层。

    ```
    .layer(0, new RBM.Builder(RBM.HiddenUnit.RECTIFIED,
        RBM.VisibleUnit.GAUSSIAN)
    ```

```
    .nIn(numRows * numColumns)

    .nOut(3)
    .weightInit(WeightInit.XAVIER)
     .k(1)
    .activation("relu")
    .lossFunction(LossFunctions.LossFunction.RMSE_XENT)
    .updater(Updater.ADAGRAD)
     .dropOut(0.5)
    .build()
 )
```

- 第一行中的 0 为层索引。
- k() 为对比散度（contrastive divergence）。
- 这里我们并没有使用二值 RBM（binary RBM），因为 Iris 数据是浮点数，这使得我们无法使用它。所以我们使用了 RBM.VisibleUnit.GAUSSIAN，让模型可以处理连续值。
- 使用 Updater.ADAGRAD 优化学习率。

12. 然后，添加如下方法到上一步的代码中。这段代码用来为神经网络创建层 1。

```
.layer(1, new
    OutputLayer.Builder(LossFunctions.LossFunction.MCXENT)

      .nIn(3)

      .nOut(outputNum)
      .activation("softmax")
      .build()
 ) .build();
```

13. 创建模型并进行初始化。

```
MultiLayerNetwork model = new MultiLayerNetwork(conf);
model.init();
```

14. 配置好模型之后，进行训练。

```
model.setListeners(Arrays.asList((IterationListener) new
```

```
                    ScoreIterationListener(listenerFreq)));
            log.info("Train model....");
            model.fit(train);
```

15. 使用如下代码段对权重进行评估。

    ```
    log.info("Evaluate weights....");
    for(org.deeplearning4j.nn.api.Layer layer : model.getLayers())
    {
        INDArray w =
         layer.getParam(DefaultParamInitializer.WEIGHT_KEY);
        log.info("Weights: " + w);
    }
    ```

16. 最后，评估模型。

    ```
    log.info("Evaluate model....");
    Evaluation eval = new Evaluation(outputNum);
    INDArray output = model.output(test.getFeatureMatrix());
    for (int i = 0; i < output.rows(); i++) {
        String actual =
            test.getLabels().getRow(i).toString().trim();
        String predicted = output.getRow(i).toString().trim();
        log.info("actual " + actual + " vs predicted " +
            predicted);
    }
    eval.eval(test.getLabels(), output);
    log.info(eval.stats());
    ```

17. 这段代码输出如下：

    ```
    =========================Scores===================================
    Accuracy:   0.8333
    Precision:  1
    Recall:     0.8333
    F1 Score:   0.9090909090909091
    ```

18. 最后，关闭 main 方法与类。

    ```
        }
    }
    ```

工作原理

1. 在 Eclipse 中的项目名上单击鼠标右键，选择 New，然后选择 Package，输入的包

名为 `deepbelief.chap8.science.data`，单击 Finish 按钮。

2. 创建好包之后，右键单击包名，依次选择 **New|Class**，输入的类名为 `DBNIrisExample`，单击 Finish 按钮。

在编辑器中，复制与粘贴如下代码。

```java
package deepbelief.chap8.science.data;


import org.deeplearning4j.datasets.iterator.DataSetIterator;
import org.deeplearning4j.datasets.iterator.impl.IrisDataSetIterator;
import org.deeplearning4j.eval.Evaluation;
import org.deeplearning4j.nn.api.OptimizationAlgorithm;
import org.deeplearning4j.nn.conf.MultiLayerConfiguration;
import org.deeplearning4j.nn.conf.NeuralNetConfiguration;
import org.deeplearning4j.nn.conf.Updater;
import org.deeplearning4j.nn.conf.layers.OutputLayer;
import org.deeplearning4j.nn.conf.layers.RBM;
import org.deeplearning4j.nn.multilayer.MultiLayerNetwork;
import org.deeplearning4j.nn.params.DefaultParamInitializer;
import org.deeplearning4j.nn.weights.WeightInit;
import org.deeplearning4j.optimize.api.IterationListener;
import org.deeplearning4j.optimize.listeners.ScoreIterationListener;
import org.nd4j.linalg.api.ndarray.INDArray;
import org.nd4j.linalg.dataset.DataSet;
import org.nd4j.linalg.dataset.SplitTestAndTrain;
import org.nd4j.linalg.factory.Nd4j;
import org.nd4j.linalg.lossfunctions.LossFunctions;
import org.slf4j.Logger;
import org.slf4j.LoggerFactory;
import java.util.Arrays;
import java.util.Random;

public class DBNIrisExample {

    private static Logger log =
      LoggerFactory.getLogger(DBNIrisExample.class);

    public static void main(String[] args) throws Exception {
        Nd4j.MAX_SLICES_TO_PRINT = -1;
        Nd4j.MAX_ELEMENTS_PER_SLICE = -1;
```

```java
final int numRows = 4;
final int numColumns = 1;
int outputNum = 3;
int numSamples = 150;
int batchSize = 150;
int iterations = 5;
int splitTrainNum = (int) (batchSize * .8);
int seed = 123;
int listenerFreq = 1;
log.info("Load data....");
DataSetIterator iter = new IrisDataSetIterator(batchSize,
  numSamples);
DataSet next = iter.next();
next.normalizeZeroMeanZeroUnitVariance();

log.info("Split data....");
SplitTestAndTrain testAndTrain =
   next.splitTestAndTrain(splitTrainNum, new Random(seed));
DataSet train = testAndTrain.getTrain();
DataSet test = testAndTrain.getTest();
Nd4j.ENFORCE_NUMERICAL_STABILITY = true;

log.info("Build model....");
MultiLayerConfiguration conf = new
  NeuralNetConfiguration.Builder()
    .seed(seed)
    .iterations(iterations)
    .learningRate(1e-6f)
    .optimizationAlgo(OptimizationAlgorithm.CONJUGATE_GRADIENT)
    .l1(1e-1).regularization(true).l2(2e-4)
    .useDropConnect(true)
    .list(2)
    .layer(0, new RBM.Builder(RBM.HiddenUnit.RECTIFIED,
       RBM.VisibleUnit.GAUSSIAN)
    .nIn(numRows * numColumns)
    .nOut(3)
    .weightInit(WeightInit.XAVIER)
    .k(1)
    .activation("relu")
    .lossFunction(LossFunctions.LossFunction.RMSE_XENT)
    .updater(Updater.ADAGRAD)
    .dropOut(0.5)
    .build()
```

```
        )
        .layer(1, new
            OutputLayer.Builder(LossFunctions.LossFunction.MCXENT)
          .nIn(3)
          .nOut(outputNum)
          .activation("softmax")
          .build()
    )
    .build();
MultiLayerNetwork model = new MultiLayerNetwork(conf);
model.init();

model.setListeners(Arrays.asList((IterationListener) new
    ScoreIterationListener(listenerFreq)));
    log.info("Train model....");
    model.fit(train);

    log.info("Evaluate weights....");
    for(org.deeplearning4j.nn.api.Layer layer : model.getLayers())
    {
        INDArray w =
        layer.getParam(DefaultParamInitializer.WEIGHT_KEY);
        log.info("Weights: " + w);
    }

    log.info("Evaluate model....");
    Evaluation eval = new Evaluation(outputNum);
    INDArray output = model.output(test.getFeatureMatrix());

    for (int i = 0; i < output.rows(); i++) {
        String actual =
           test.getLabels().getRow(i).toString().trim();
        String predicted = output.getRow(i).toString().trim();
        log.info("actual " + actual + " vs predicted " +
           predicted);
    }

    eval.eval(test.getLabels(), output);
    log.info(eval.stats());
  }
}
```

8.4 使用 DL4j 创建深度自动编码器

深度自动编码器是一种深度神经网络，由两个对称的深度信念网络组成。这些网络通常有两个单独的 4 个或 5 个浅层（限制玻尔兹曼机），用来表示编码与解码半网络。本节中，我们将创建一个深度自动编码器，它由一个输入层、4 个解码层、4 个编码层与一个输出层组成。在此过程中，我们将使用一个非常有名的数据集，即 MNIST 数据集。

要想学习更多关于 MNIST 的内容，请访问 http://yann.lecun.com/exdb/mnist/。若想了解更多有关深度自动编码器的内容，请访问 https://deeplearning4j.org/deepautoencoder。为了完成命令，关闭所有打开的窗口。在菜单栏中，依次单击 **File|New|Other...**。直到你到达如下窗口。在该窗口中，在 Group Id 与 Artifact Id 中填入如下名称，单击 Finish。

操作步骤

1. 创建一个名为 DeepAutoEncoderExample 的类。

    ```
    public class DeepAutoEncoderExample {
    ```

2. 在整个代码中，我们要记录信息。所以，为类创建一个 Logger。

    ```
    private static Logger log =
        LoggerFactory.getLogger(DeepAutoEncoderExample.class);
    ```

3. 开始编写 main 方法。

    ```
    public static void main(String[] args) throws Exception {
    ```

4. 在 main 方法的最开始部分，定义几个要修改或配置的参数。

    ```
    final int numRows = 28;
    final int numColumns = 28;
    int seed = 123;
    int numSamples = MnistDataFetcher.NUM_EXAMPLES;
    int batchSize = 1000;
    int iterations = 1;
    ```

```
int listenerFreq = iterations/5;
```

- 行数与列数都设置为 28，因为 MNIST 数据集中图像尺寸为 28 像素×28 像素。
- 设置随机种子为 123。
- `numSamples` 是示例数据集中总样本数。
- `batchSize` 设置为 1 000，即每次使用 1 000 个数据样本。
- `listenerFreq` 是指查看注册到进程时损失函数值的频率。

5. 然后，使用指定的批大小与样本数信息，加载 MNIST 数据点。

```
log.info("Load data....");
DataSetIterator iter = new
    MnistDataSetIterator(batchSize,numSamples,true);
```

6. 接下来，开始配置神经网络。首先，使用指定的种子、迭代次数创建一个多层神经网络，并把线性梯度下降设为优化算法。总共要设置 10 个层：一个输入层、4 个编码层、一个解码层、一个输出层。

```
log.info("Build model....");
MultiLayerConfiguration conf = new
 NeuralNetConfiguration.Builder()
.seed(seed)
.iterations(iterations)
.optimizationAlgo(OptimizationAlgorithm.LINE_GRADIENT_DESCENT)
.list(10)
```

7. 然后，把如下代码添加到上一步代码中，这些代码用来创建 10 个层，带有反向传播设置。

```
.layer(0, new RBM.Builder().nIn(numRows *
    numColumns).nOut(1000).lossFunction
      (LossFunctions.LossFunction.RMSE_XENT).build())
.layer(1, new RBM.Builder().nIn(1000).nOut(500).lossFunction
      (LossFunctions.LossFunction.RMSE_XENT).build())
.layer(2, new RBM.Builder().nIn(500).nOut(250).lossFunction
      (LossFunctions.LossFunction.RMSE_XENT).build())
.layer(3, new RBM.Builder().nIn(250).nOut(100).lossFunction
      (LossFunctions.LossFunction.RMSE_XENT).build())
.layer(4, new RBM.Builder().nIn(100).nOut(30).lossFunction
```

```
            (LossFunctions.LossFunction.RMSE_XENT).build())
        //编码停止
    .layer(5, new RBM.Builder().nIn(30).nOut(100).lossFunction
            (LossFunctions.LossFunction.RMSE_XENT).build())
        //解码开始
    .layer(6, new RBM.Builder().nIn(100).nOut(250).lossFunction
            (LossFunctions.LossFunction.RMSE_XENT).build())
    .layer(7, new RBM.Builder().nIn(250).nOut(500).lossFunction
            (LossFunctions.LossFunction.RMSE_XENT).build())
    .layer(8, new RBM.Builder().nIn(500).nOut(1000).lossFunction
            (LossFunctions.LossFunction.RMSE_XENT).build())
    .layer(9, new OutputLayer.Builder(LossFunctions.
            LossFunction.RMSE_XENT).nIn(1000).nOut(numRows*numColumns).
               build())
    .pretrain(true).backprop(true) .build();
```

8. 配置好模型之后，接着对其初始化。

    ```
    MultiLayerNetwork model = new MultiLayerNetwork(conf);
    model.init();
    ```

9. 对模型进行训练。

    ```
    model.setListeners(Arrays.asList((IterationListener) new
    ScoreIterationListener(listenerFreq)));
      log.info("Train model....");
      while(iter.hasNext()) {
     DataSet next = iter.next();
      model.fit(new
        DataSet(next.getFeatureMatrix(),next.getFeatureMatrix()));
    }
    ```

10. 最后，关闭 main 方法与类

    ```
        }
        }
    ```

工作原理

1. 在 Eclipse 中的项目名上单击鼠标右键，选择 New，然后选择 Package，输入的包名为 deepbelief.chap8.science.data，单击 Finish 按钮。

2. 创建好包之后，右键单击包名，依次选择 **New|Class**，输入的类名为 `DeepAutoEncoder-Example`，单击 **Finish** 按钮。

在编辑器中，复制与粘贴如下代码。

```java
package deepbelief.chap8.science.data;
import org.deeplearning4j.datasets.fetchers.MnistDataFetcher;
import org.deeplearning4j.datasets.iterator.impl.MnistDataSetIterator;
import org.deeplearning4j.nn.api.OptimizationAlgorithm;
import org.deeplearning4j.nn.conf.MultiLayerConfiguration;
import org.deeplearning4j.nn.conf.NeuralNetConfiguration;
import org.deeplearning4j.nn.conf.layers.OutputLayer;
import org.deeplearning4j.nn.conf.layers.RBM;
import org.deeplearning4j.nn.multilayer.MultiLayerNetwork;
import org.deeplearning4j.optimize.api.IterationListener;
import org.deeplearning4j.optimize.listeners.ScoreIterationListener;
import org.nd4j.linalg.dataset.DataSet;
import org.nd4j.linalg.dataset.api.iterator.DataSetIterator;
import org.nd4j.linalg.lossfunctions.LossFunctions;
import org.slf4j.Logger;
import org.slf4j.LoggerFactory;
import java.util.Arrays;

public class DeepAutoEncoderExample {
    private static Logger log =
        LoggerFactory.getLogger(DeepAutoEncoderExample.class);

    public static void main(String[] args) throws Exception {
        final int numRows = 28;
        final int numColumns = 28;
        int seed = 123;
        int numSamples = MnistDataFetcher.NUM_EXAMPLES;
        int batchSize = 1000;
        int iterations = 1;
        int listenerFreq = iterations/5;

        log.info("Load data....");
        DataSetIterator iter = new
          MnistDataSetIterator(batchSize,numSamples,true);

        log.info("Build model....");
        MultiLayerConfiguration conf = new
          NeuralNetConfiguration.Builder()
                .seed(seed)
```

```
                .iterations(iterations)
        .optimizationAlgo(OptimizationAlgorithm.LINE_GRADIENT_DESCENT)
                .list(10)
        .layer(0, new RBM.Builder().nIn(numRows *
            numColumns).nOut(1000).lossFunction
               (LossFunctions.LossFunction.RMSE_XENT).build())
        .layer(1, new RBM.Builder().nIn(1000).nOut(500).lossFunction
               (LossFunctions.LossFunction.RMSE_XENT).build())
        .layer(2, new RBM.Builder().nIn(500).nOut(250).lossFunction
               (LossFunctions.LossFunction.RMSE_XENT).build())
        .layer(3, new RBM.Builder().nIn(250).nOut(100).lossFunction
               (LossFunctions.LossFunction.RMSE_XENT).build())
        .layer(4, new RBM.Builder().nIn(100).nOut(30).lossFunction
               (LossFunctions.LossFunction.RMSE_XENT).build())
               //编码停止
        .layer(5, new RBM.Builder().nIn(30).nOut(100).lossFunction
               (LossFunctions.LossFunction.RMSE_XENT).build())
               //解码开始
        .layer(6, new RBM.Builder().nIn(100).nOut(250).lossFunction
               (LossFunctions.LossFunction.RMSE_XENT).build())
        .layer(7, new RBM.Builder().nIn(250).nOut(500).lossFunction
               (LossFunctions.LossFunction.RMSE_XENT).build())
        .layer(8, new RBM.Builder().nIn(500).nOut(1000).lossFunction
               (LossFunctions.LossFunction.RMSE_XENT).build())
        .layer(9, new OutputLayer.Builder(LossFunctions.
            LossFunction.RMSE_XENT).nIn(1000).nOut(numRows*numColumns).
               build())
        .pretrain(true).backprop(true) .build();

        MultiLayerNetwork model = new MultiLayerNetwork(conf);
        model.init();

        model.setListeners(Arrays.asList((IterationListener) new
            ScoreIterationListener(listenerFreq)));

        log.info("Train model....");
        while(iter.hasNext()) {
            DataSet next = iter.next();
            model.fit(new
          DataSet(next.getFeatureMatrix(),next.getFeatureMatrix()));
        }
    }
}
```

第 9 章
数据可视化

本章涵盖如下内容：

- 绘制 2D 正弦；
- 绘制直方图；
- 绘制柱状图；
- 绘制箱线图或箱须图；
- 绘制散点图；
- 绘制甜圈图；
- 绘制面积图。

9.1 简介

在数据科学界，数据可视化正变得越来越流行，它借助点、线、条形把隐藏在数据之下的信息以视觉化方式呈现出来。通过数据可视化，不仅可以把信息传递给数据科学家，还可以把信息直观地呈现给那些不懂（或懂得不多）底层数据分布与数据性质的观众。在许多场合下，管理人、赌金保管人、业务主管会采用数据可视化技术来进行决策或理解变化趋势。

本章分了 8 个小节来讲解数据可视化，所涉及的可视化图形有正弦图、直观图、柱状图、箱线图、散点图、甜圈图或饼图、面积图。讲解中，我们不会对这些图形的背景知识、优缺点、用法讲太多，但会进行一点简单的介绍。取而代之，我们会把讲解重点放在如何使用 Java 库来创建这些可视化图形上来。

本章，我们会用到一个 Java 数据可视化库，它叫 GRAL，是 GRAphing Library 的缩写。本章之所以把 GRAL 用作数据可视化库，原因有如下几点。

- 包含的类全面、综合。

- 包含平滑、缩放、统计、直方图等数据处理功能。

- 支持绘制广受数据科学家喜爱的图形，包括：xy/散点图、气泡图、折线图、面积图、柱状图、饼图、甜圈图、箱线图、点阵图。

- 支持图例显示。

- 支持多种文件格式作为数据源或数据接收器，比如 CSV、位图图像数据、音频文件数据。

- 支持导出位图与矢量文件格式的图形，比如 PNG、GIF、JPEG、EPS、PDF、SVG。

- 占用内存少（大约 300k）。

感兴趣的读者，建议前往如下地址查看多个 Java 数据可视化库的比较：https://github.com/eseifert/gral/wiki/comparison。

9.2 绘制 2D 正弦曲线

本节中，我们将学习使用免费的 Java 图形库——GRAL（GRAphing Library）来绘制 2D 正弦曲线。在许多情况下，正弦曲线对数据科学家特别有用，因为这种三角函数的图形可以用来对数据的波动进行建模（比如，使用温度数据创建一个模型，用来预测一年当中当地适于观光的时间）。

准备工作

1. 为了在项目中使用 GRAL，需要先下载 GRAL JAR 文件，并将其作为外部 Java 库添加到项目中。前往 http://trac.erichseifert.de/gral/wiki/Download，在页面的 "Legacy versions" 部分，下载 GRAL 0.10 版本，即下载名为 gral-core-0.10.zip 的 zip 文件（见图 9-1）。

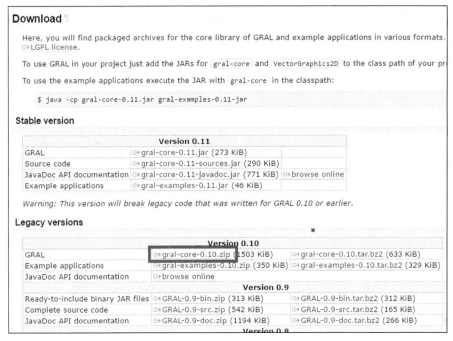

图 9-1

下载完成后,进行解压缩,将看到如下文件与文件夹。如图 9-2 所示,有一个名为 `lib` 的文件夹,它即是我们要用的文件夹。

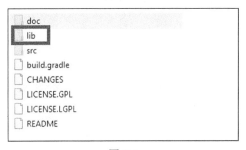

图 9-2

2. 进入 `lib` 文件夹,里面包含 `gral-core-0.10` 与 `VectorGraphics2D-0.9.1` 两个 Jar 文件。这里,我们只使用图 9-3 所示的 `gral-core-0.10.jar` 文件。

图 9-3

3. 如图 9-4 所示，在 Eclipse 项目中，把 gral-core-0.10.jar 这个 JAR 文件作为外部库文件添加到项目中。

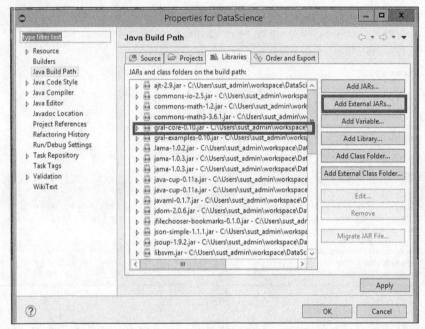

图 9-4

4. 接下来，就该编写绘制正弦曲线的代码了。

操作步骤

1. 首先，创建一个名为 SineGraph 的 Java 类，它继续了 JFrame，因为我们要把数据图形绘制到一个 JFrame 中。

   ```
   public class SineGraph extends JFrame {
   ```

2. 接着，声明一个 serialVersionUID 作为类变量。

   ```
   private static final long serialVersionUID = 1L;
   ```

3. 我们可以把 serialVersionUID 用作 Serializable 类的版本控制。如果没有显式声明 serialVersionUID，JVM 会自动为我们做。相关细节已超出本书讨论范围，感兴趣的读者，请前往 http://docs.oracle.com/javase/1.5.0/docs/api/java/io/Serializable.html 页面了解更多的细节。

4. 然后，为这个类创建一个构造函数。在构造函数中定义关闭窗体（frame）时的行为，用来绘制正弦曲线的窗体尺寸，还要在 for 循环中创建数据表。本例中，我们会看到一条真正的正弦曲线。请注意，使用你的真实数据可能无法生成一条完美的正弦曲线。

   ```
   public SineGraph() throws FileNotFoundException, IOException {
   ```

5. 定义关闭窗体时的默认行为。

   ```
   setDefaultCloseOperation(EXIT_ON_CLOSE);
   ```

6. 设置窗体大小。

   ```
   setSize(1600, 1400);
   ```

7. 使用 for 循环与 sin() 函数生成一系列的 x 与 y 值，然后把它们放入一个数据表中。

   ```
   DataTable data = new DataTable(Double.class, Double.class);
    for (double x = -5.0; x <= 5.0; x+=0.25) {
    double y = 5.0*Math.sin(x);
    ata.add(x, y);
    }
   ```

8. 接下来，我们将使用 GRAL 的 XYPlot 类来绘制正弦曲线。首先，创建一个 XYPlot 对象，并把上一步中生成的数据作为参数传递给它。

   ```
   XYPlot plot = new XYPlot(data);
   ```

9. 把图形设置到交互面板。

   ```
   XYPlot plot = new XYPlot(data);
   ```

10. 为了渲染图形，先创建一个 2D LineRenderer。然后把这个 2D LineRenderer 与数据添加到 XYPlot 对象。

    ```
    LineRenderer lines = new DefaultLineRenderer2D();
    plot.setLineRenderer(data, lines);
    ```

11. 在 GRAL 中，我们可以使用 Color 类绘制彩色图形。

    ```
    Color color = new Color(0.0f, 0.0f, 0.0f);
    ```

12. 在创建 Color 类的对象时，我们需要为 Color 类的构造函数提供红、绿、蓝颜色值。在上一步中，我们把红、绿、蓝的颜色值全部设为 0，即将颜色设置为黑色，用以绘制图形。

13. 为点与线条设置颜色。

    ```
    plot.getPointRenderer(data).setColor(color);
    plot.getLineRenderer(data).setColor(color);
    ```

14. 关闭构造函数。

    ```
    }
    ```

15. 为了运行程序，编写如下 main() 方法。

    ```
    public static void main(String[] args) {
    SineGraph frame = null;
    try {
     frame = new SineGraph();
    } catch (IOException e) {
    }
    frame.setVisible(true);
    }
    ```

示例的完整代码整理如下：

```
import java.awt.Color;
import java.io.FileNotFoundException;
import java.io.IOException;
import javax.swing.JFrame;
import de.erichseifert.gral.data.DataTable;
import de.erichseifert.gral.plots.XYPlot;
import de.erichseifert.gral.plots.lines.DefaultLineRenderer2D;
import de.erichseifert.gral.plots.lines.LineRenderer;
import de.erichseifert.gral.ui.InteractivePanel;

public class SineGraph extends JFrame {
   private static final long serialVersionUID = 1L;

   public SineGraph() throws FileNotFoundException, IOException {
      setDefaultCloseOperation(EXIT_ON_CLOSE);
      setSize(1600, 1400);

      DataTable data = new DataTable(Double.class, Double.class);
```

```
        for (double x = -5.0; x <= 5.0; x+=0.25) {
            double y = 5.0*Math.sin(x);
            data.add(x, y);
        }
        XYPlot plot = new XYPlot(data);
        getContentPane().add(new InteractivePanel(plot));
        LineRenderer lines = new DefaultLineRenderer2D();
        plot.setLineRenderer(data, lines);
        Color color = new Color(0.0f, 0.3f, 1.0f);
        plot.getPointRenderer(data).setColor(color);
        plot.getLineRenderer(data).setColor(color);
    }
    public static void main(String[] args) {
        SineGraph frame = null;
        try {
            frame = new SineGraph();
        } catch (IOException e) {
        }
        frame.setVisible(true);
    }
}
```

运行上面代码，绘制正弦曲线如图 9-5 所示。

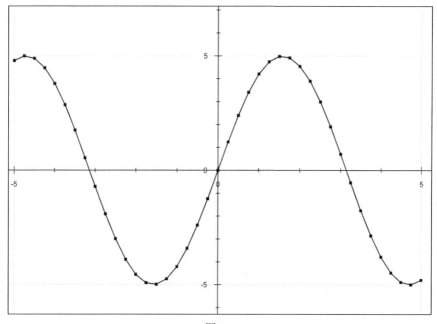

图 9-5

9.3 绘制直方图

在探索一系列连续数据的频率分布时，直方图是一种很常用的方法。在直方图中，数据科学家通常会把 x 轴定义为定量变量，把 y 轴定义为该变量的频率。直方图拥有如下几个主要特征，这些特征让直方图非常有用。

- 只能绘制数值数据。
- 可以轻松地绘制大数据集。
- x 轴通常用作 bins 或定量变量的间隔。

本节，我们将学习如何使用 GRAL 绘制直方图。

准备工作

1. 为了使用 GARL 绘制直方图，我们需要随该库一起提供的示例程序，这些示例是 Jar 文件。你可以从 http://trac.erichseifert.de/gral/wiki/Download 页面把图 9-6 所示的 gral-examples-0.10.zip 文件下载到本地磁盘中，然后进行解压缩。

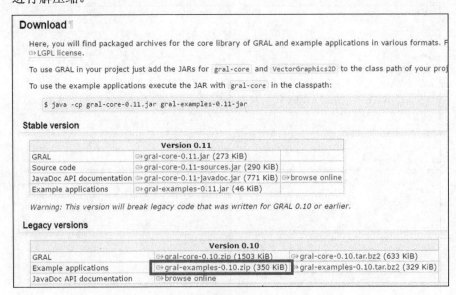

图 9-6

2. 下载好 Zip 文件之后，进行解压缩，你将看到如图 9-7 所示的目录结构，其中的 lib

文件夹是我们所感兴趣的。

图 9-7

3. 在 `lib` 文件夹中，包含 `gral-core-0.10`、`gralexamples-0.10`、`VectorGraphics2D-0.9.1` 3 个 Jar 文件。如图 9-8 所示，第一个文件在本章第一节中已经用过。本节，我们将使用第二个 Jar 文件，即 `gralexamples-0.10` 文件。

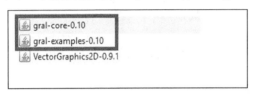

图 9-8

4. 把图 9-9 中的 `gral-core-0.10` 和 `gralexamples-0.10` 两个 Jar 文件作为外部库添加到项目中。

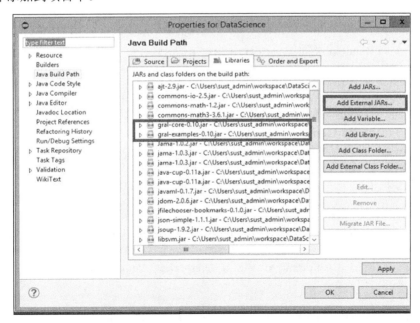

图 9-9

5. 接下来，我们就可以使用 GRAL 示例包中的程序来绘制直方图了。在下载的示例包中 `gral-examples-0.10\gralexamples-0.10\src\main\java\de\erichseifert\gral\examples\ barplot` 中，你可以看到我们要用的程序。

操作步骤

1. 创建一个名为 `HistogramPlot` 类，它继承自 `ExamplePanel` 类。然后创建一个 `serialVersionUID`。

    ```
    public class HistogramPlot extends ExamplePanel {
    private static final long serialVersionUID =
        4458280577519421950L;
    ```

2. 本例中，我们将为 1 000 个样本数据点创建直方图。

    ```
    private static final int SAMPLE_COUNT = 1000;
    ```

3. 创建 `HistogramPlot` 类的构造函数。

    ```
    public HistogramPlot() {
    ```

4. 随机生成 1 000 个样本数据点。这些数据点遵从高斯分布，因为创建它们时我们使用了 Java `Random` 类的 `random.nextGaussian()` 方法。

    ```
    Random random = new Random();
    DataTable data = new DataTable(Double.class);
     for (int i = 0; i < SAMPLE_COUNT; i++) {
        data.add(random.nextGaussian());
     }
    ```

5. 根据数据创建一个 `histogram` 变量，然后为绘图创建另一个维度。

    ```
    Histogram1D histogram = new Histogram1D(data,
     Orientation.VERTICAL,new Number[] {-4.0, -3.2, -2.4, -1.6,
      -0.8, 0.0, 0.8, 1.6, 2.4, 3.2, 4.0});
    DataSource histogram2d = new EnumeratedData(histogram, (-4.0 +
      -3.2)/2.0, 0.8);
    ```

6. 数组中的数值是直方图在 x 轴方向上的刻度，如图 9-10 所示。

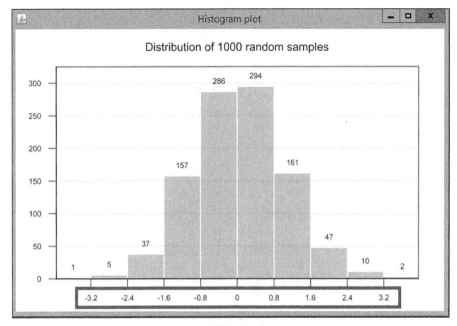

图 9-10

7. 如你所见，直方图是一种条形图。因此，我们要先创建一个 BarPlot 对象，并且把直方图信息提供给它。

   ```
   BarPlot plot = new BarPlot(histogram2d);
   ```

8. 接下来，为绘图区域设置格式。

9. 设置直方图在窗体中的坐标。

   ```
   plot.setInsets(new Insets2D.Double(20.0, 65.0, 50.0, 40.0));
   ```

10. 为直方图设置标题。

    ```
    plot.getTitle().setText(
    String.format("Distribution of %d random samples",
      data.getRowCount()));
    ```

11. 为直方图的长条设置宽度。

    ```
    plot.setBarWidth(0.78);
    ```

12. 为 *x* 轴设置格式。如果你熟悉 Microsoft Excel，那么你应该知道对于一个指定的坐标轴会有一些选项用来设置刻度标志的对齐方式与间隔。如果用户想查看坐标轴上的小刻度标志，他们可以自己进行选择。幸运的是，GRAL 为我们提供了这个功能，这会让我们绘出的图形在科学界更具吸引力。

13. 为 *x* 轴设置刻度标志对齐方式。请注意，getAxisRenderer()方法的参数，它针对的是 *x* 轴。

    ```
    plot.getAxisRenderer(BarPlot.AXIS_X).setTickAlignment(0.0);
    ```

14. 设置刻度间隔。

    ```
    plot.getAxisRenderer(BarPlot.AXIS_X).setTickSpacing(0.8);
    ```

15. 最后，把较小刻度设置为不可见，把它们隐藏起来。

    ```
    plot.getAxisRenderer(BarPlot.AXIS_X).setMinorTicksVisible(false);
    ```

16. 接下来，设置 *y* 轴。本例中，我们先定义直方图中的长条可伸展的高度范围。

    ```
    plot.getAxis(BarPlot.AXIS_Y).setRange(0.0,
    MathUtils.ceil(histogram.getStatistics().get(Statistics.MAX)*1.1,
    25.0));
    ```

17. 然后，设置刻度对齐方式、间隔，以及较小刻度的可视性，参考前面我们为 *x* 轴所进行的设置。

    ```
    plot.getAxisRenderer(BarPlot.AXIS_Y).setTickAlignment(0.0);
    plot.getAxisRenderer(BarPlot.AXIS_Y).setMinorTicksVisible(false);
    plot.getAxisRenderer(BarPlot.AXIS_Y).setIntersection(-4.4);
    ```

18. 接着，为长条设置格式。先为长条设置颜色，再配置直方图，使其在长条的顶部把频率值显示出来。

    ```
    plot.getPointRenderer(histogram2d).setColor(
    GraphicsUtils.deriveWithAlpha(COLOR1, 128));
    plot.getPointRenderer(histogram2d).setValueVisible(true);
    ```

19. 最后，把图形添加到 Swing 组件。

    ```
    InteractivePanel panel = new InteractivePanel(plot);
    panel.setPannable(false);
    ```

```
        panel.setZoomable(false);
        add(panel);
```

20. 关闭构造函数。

```
    }
```

21. 此外，我们还需要实现 `ExamplePanel` 类中的所有方法。为了简单起见，这里我们只实现 `getTitle()` 与 `getDescription()` 两个方法。

```
    @Override
     public String getTitle() {
         return "Histogram plot";
     }
    @Override
       public String getDescription() {
         return String.format("Histogram of %d samples",
           SAMPLE_COUNT);
       }
```

22. `HistogramPlot` 类的 main 方法如下：

```
    public static void main(String[] args) {
    new HistogramPlot().showInFrame();
    }
```

23. 最后，关闭类。

```
    }
```

24. 完整的代码整理如下：

```
import java.util.Random;
import de.erichseifert.gral.data.DataSource;
import de.erichseifert.gral.data.DataTable;
import de.erichseifert.gral.data.EnumeratedData;
import de.erichseifert.gral.data.statistics.Histogram1D;
import de.erichseifert.gral.data.statistics.Statistics;
import de.erichseifert.gral.examples.ExamplePanel;
import de.erichseifert.gral.plots.BarPlot;
import de.erichseifert.gral.ui.InteractivePanel;
```

```java
import de.erichseifert.gral.util.GraphicsUtils;
import de.erichseifert.gral.util.Insets2D;
import de.erichseifert.gral.util.MathUtils;
import de.erichseifert.gral.util.Orientation;

public class HistogramPlot extends ExamplePanel {
    /**序列化版本id*/
    private static final long serialVersionUID = 4458280577519421950L;

    private static final int SAMPLE_COUNT = 1000;

    //@SuppressWarnings("unchecked")
    public HistogramPlot() {
        //生成示例数据
        Random random = new Random();
        DataTable data = new DataTable(Double.class);
        for (int i = 0; i < SAMPLE_COUNT; i++) {
            data.add(random.nextGaussian());
        }

        //根据数据生成直方图
        Histogram1D histogram = new Histogram1D(data,
         Orientation.VERTICAL, new Number[] {-4.0, -3.2, -2.4, -1.6,
            -0.8, 0.0, 0.8, 1.6, 2.4, 3.2, 4.0});
        //为图形创建另一个维度(x轴)
        DataSource histogram2d = new EnumeratedData(histogram, (-4.0 +
            -3.2)/2.0, 0.8);

        //新建条形图
        BarPlot plot = new BarPlot(histogram2d);

        //设置图形格式
        plot.setInsets(new Insets2D.Double(20.0, 65.0, 50.0, 40.0));
        plot.getTitle().setText(
            String.format("Distribution of %d random samples",
                data.getRowCount()));
        plot.setBarWidth(0.78);

        //设置x轴格式
        plot.getAxisRenderer(BarPlot.AXIS_X).setTickAlignment(0.0);
```

```
        plot.getAxisRenderer(BarPlot.AXIS_X).setTickSpacing(0.8);
        plot.getAxisRenderer(BarPlot.AXIS_X).setMinorTicksVisible(false);
        //设置 y 轴格式
        plot.getAxis(BarPlot.AXIS_Y).setRange(0.0,
            MathUtils.ceil(histogram.getStatistics().
               get(Statistics.MAX)*1.1, 25.0));
        plot.getAxisRenderer(BarPlot.AXIS_Y).setTickAlignment(0.0);
        plot.getAxisRenderer(BarPlot.AXIS_Y).setMinorTicksVisible(false);
        plot.getAxisRenderer(BarPlot.AXIS_Y).setIntersection(-4.4);

        //设置长条格式
        plot.getPointRenderer(histogram2d).setColor(
            GraphicsUtils.deriveWithAlpha(COLOR1, 128));
        plot.getPointRenderer(histogram2d).setValueVisible(true);

        //把图形添加到 Swing 组件
        InteractivePanel panel = new InteractivePanel(plot);
        panel.setPannable(false);
        panel.setZoomable(false);
        add(panel);
    }
    @Override
    public String getTitle() {
        return "Histogram plot";
    }

    @Override
    public String getDescription() {
        return String.format("Histogram of %d samples", SAMPLE_COUNT);
    }

    public static void main(String[] args) {
        new HistogramPlot().showInFrame();
    }
}
```

9.4 绘制条形图

如图 9-11 所示,条形图是数据科学家最常用的图形类型。使用 GRAL 可以很容易地绘制出条形图。在本节中，我们将学习使用 GRAL 绘制条形图的方法。

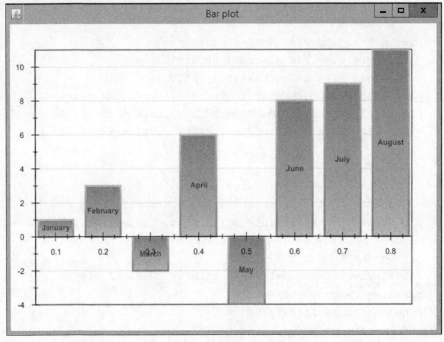

图 9-11

准备工作

1. 为了使用 GARL 绘制条形图，我们需要随该库一起提供的示例程序，这些示例是 Jar 文件。你可以从 http://trac.erichseifert.de/gral/wiki/Download 页面把 gral-examples-0.10.zip 下载到本地磁盘中，然后进行解压缩。

2. 下载好 Zip 文件之后，进行解压缩，在解压得到的目录结构（请参考 9.2 节中的准备部分）中，包含一个 lib 文件夹，该文件夹正是我们所感兴趣的。

3. 在 lib 文件夹中，包含 gral-core-0.10、gralexamples-0.10、VectorGraphics2D-0.9.1 3 个 Jar 文件。本节，我们将使用前两个 Jar 文件，即 gral-core-0.10 与 gralexamples-0.10。

4. 把 gral-core-0.10 与 gralexamples-0.10 两个 Jar 文件作为外部库添加到项目中。

接下来，我们就可以使用 GRAL 示例包中的程序来绘制条形图了。在下载示例包的 gral-examples-0.10\gralexamples-0.10\src\main\java\de\erichseifer t\gral\examples\barplot 中，你可以看到我们要用的程序。

操作步骤

1. 创建一个名为 `SimpleBarPlot` 的类。类似于上一节，这个类也继承了 GRAL 库的 `ExamplePanel` 类。

   ```
   publicclassSimpleBarPlotextendsExamplePanel {
   ```

2. 创建一个序列版本 UID。

   ```
   privatestaticfinallong serialVersionUID =-2793954497895054530L;
   ```

3. 编写 `SimpleBarPlot` 类的构造函数。

   ```
   publicSimpleBarPlot() {
   ```

4. 首先，创建示例数据。在本节开始部分的条形图中，每个长条有 3 个值，分别是 x 轴的值、y 轴的值与条形图名称。比如，在第一个长条中，x 轴的值为 0.1，y 轴的值为 1，名称为 January。我们可以采用如下方式为所有长条创建数据点。

   ```
   DataTable data = new DataTable(Double.class, Integer.class,
    String.class);
   data.add(0.1, 1, "January");
   data.add(0.2, 3, "February");
   data.add(0.3, -2, "March");
   data.add(0.4, 6, "April");
   data.add(0.5, -4, "May");
   data.add(0.6, 8, "June");
   data.add(0.7, 9, "July");
   data.add(0.8, 11, "August");
   ```

5. `DataTable` 类的构造函数接收 3 个值，分别是 x 轴值（double）、y 轴值（integer）、长条名称（String）。

6. 其余代码用来设置条形图。

7. 新建一个长条图。

   ```
   BarPlot plot = newBarPlot(data);
   ```

8. 设置长条图的尺寸与长条宽度。

```
plot.setInsets(new Insets2D.Double(40.0, 40.0, 40.0, 40.0));
plot.setBarWidth(0.075);
```

9. 接下来，我们设置长条格式。为此，先使用数据创建一个 BarRenderer 对象。

    ```
    BarRenderer pointRenderer = (BarRenderer)
      plot.getPointRenderer(data);
    ```

10. 然后，设置长条颜色。

    ```
    pointRenderer.setColor(
      new LinearGradientPaint(0f,0f, 0f,1f,
      new float[] { 0.0f, 1.0f },
      new Color[] { COLOR1, GraphicsUtils.deriveBrighter(COLOR1) }
      )
    );
    ```

11. 接着，设置长条图的属性。

    使用如下代码，在条形图中显示数值。

    ```
    pointRenderer.setValueVisible(true);
    ```

    把数据中的第三个值设置为分值栏。

    ```
    pointRenderer.setValueColumn(2);
    ```

    把值的位置设置到中心。

    ```
    pointRenderer.setValueLocation(Location.CENTER);
    ```

    设置值的颜色。

    ```
    pointRenderer.setValueColor(GraphicsUtils.deriveDarker(COLOR1));
    ```

    开启字体的粗体显示方式。

    ```
    pointRenderer.setValueFont(Font.decode
        (null).deriveFont(Font.BOLD));
    ```

12. 把条形图添加到 Swing 组件中。

    ```
    add(newInteractivePanel(plot));
    ```

13. 关闭构造函数。

    ```
    }
    ```

14. 接下来，我们还需要实现 GRAL 库中 ExamplePanel 类的另外两个方法。

    ```
     @Override
    public String getTitle() {
     return "Bar plot";
    }
     @Override
    public String getDescription() {
        return "Bar plot with example data and color gradients";
    }
    ```

15. 编写用来运行代码的 main 方法，如下：

    ```
    public static void main(String[] args) {
    new SimpleBarPlot().showInFrame();
    }
    ```

16. 关闭类。

    ```
    }
    ```

示例的完整代码整理如下：

```
import java.awt.Color;
import java.awt.Font;
import java.awt.LinearGradientPaint;
import de.erichseifert.gral.data.DataTable;
import de.erichseifert.gral.examples.ExamplePanel;
import de.erichseifert.gral.plots.BarPlot;
import de.erichseifert.gral.plots.BarPlot.BarRenderer;
import de.erichseifert.gral.ui.InteractivePanel;
import de.erichseifert.gral.util.GraphicsUtils;
import de.erichseifert.gral.util.Insets2D;
import de.erichseifert.gral.util.Location;

public class SimpleBarPlot extends ExamplePanel {
   /** 序列化版本 id */
   private static final long serialVersionUID = -2793954497895054530L;
```

```java
@SuppressWarnings("unchecked")
public SimpleBarPlot() {
    //创建示例数据
    DataTable data = new DataTable(Double.class, Integer.class,
        String.class);
    data.add(0.1, 1, "January");
    data.add(0.2, 3, "February");
    data.add(0.3, -2, "March");
    data.add(0.4, 6, "April");
    data.add(0.5, -4, "May");
    data.add(0.6, 8, "June");
    data.add(0.7, 9, "July");
    data.add(0.8, 11, "August");

    //新建条形图
    BarPlot plot = new BarPlot(data);

    //设置图形
    plot.setInsets(new Insets2D.Double(40.0, 40.0, 40.0, 40.0));
    plot.setBarWidth(0.075);

    //设置长条
    BarRenderer pointRenderer = (BarRenderer)
      plot.getPointRenderer(data);
    pointRenderer.setColor(
      new LinearGradientPaint(0f,0f, 0f,1f,
          new float[] { 0.0f, 1.0f },
          new Color[] { COLOR1,
            GraphicsUtils.deriveBrighter(COLOR1) }
      )
    );
    /*pointRenderer.setBorderStroke(new BasicStroke(3f));
    pointRenderer.setBorderColor(
      new LinearGradientPaint(0f,0f, 0f,1f,
          new float[] { 0.0f, 1.0f },
          new Color[] { GraphicsUtils.deriveBrighter(COLOR1),
            COLOR1 }
      )
    );*/
    pointRenderer.setValueVisible(true);
    pointRenderer.setValueColumn(2);
    pointRenderer.setValueLocation(Location.CENTER);
    pointRenderer.setValueColor(GraphicsUtils.deriveDarker(COLOR1));
```

```
        pointRenderer.setValueFont(Font.decode(null).deriveFont(Font.BOLD));

        //添加图形到Swing组件
        add(new InteractivePanel(plot));
    }
    @Override
    public String getTitle() {
        return "Bar plot";
    }
    @Override
    public String getDescription() {
        return "Bar plot with example data and color gradients";
    }
    public static void main(String[] args) {
        new SimpleBarPlot().showInFrame();
    }
}
```

9.5 绘制箱线图或箱须图

对数据科学家来说，箱线图是另一个有效的可视化工具。它们可以给出一个数据分布的描述性统计信息。一个典型的箱线图包含如下数据分布信息：

- 最小值；
- 第一个四分位数；
- 中位数；
- 第三个四分位数；
- 最大值。

此外，从这些统计量中也可以得到其他值，比如四分位距，它是第三个四分位数与第一个四分位数之差。

本节，我们将学习使用 GRAL 为数据分布绘制箱线图的方法。

准备工作

1. 为了使用 GARL 绘制箱线图，我们需要随该库一起提供的示例程序，这些示例以 Jar 文件形式存在。你可以从 http://trac.erichseifert.de/gral/wiki/

Download 页面把 gral-examples-0.10.zip 下载到本地磁盘中,然后进行解压缩。

2. 下载好 Zip 文件之后,进行解压缩,在解压得到的目录结构(请参考 9.2 节中的准备部分)中,包含一个 lib 文件夹,该文件夹正是我们所感兴趣的。

3. 在 lib 文件夹中,包含 gral-core-0.10、gralexamples-0.10、VectorGraphics2D-0.9.1 3 个 Jar 文件。本节,我们将使用前两个 Jar 文件,即 gral-core-0.10 与 gralexamples-0.10。

4. 把 gral-core-0.10 与 gralexamples-0.10 两个 Jar 文件作为外部库添加到项目中。

接下来,我们就可以使用 GRAL 示例包中的程序来绘制箱线图了。在下载示例包的 gral-examples-0.10\gralexamples-0.10\src\main\java\de\erichseifert\gral\examples\boxplot 中,你可以看到我们要用的程序。成功运行本节代码后,你将看到如图 9-12 所示的箱线图。

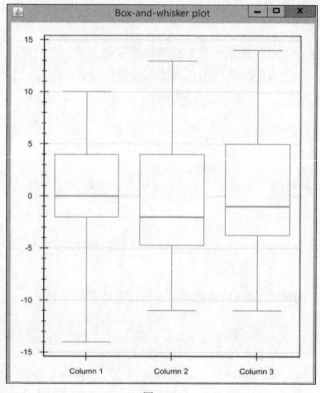

图 9-12

操作步骤

1. 首先，创建一个名为 `SimpleBoxPlot` 的类，它继承了 GRAL 库中的 `ExamplePanel` 类。在其中，声明类的序列化版本 UID。

    ```
    public class SimpleBoxPlot extends ExamplePanel {
    private static final long serialVersionUID =
        5228891435595348789L;
    ```

2. 接下来，我们将为要创建与绘制的箱线图生成 50 个随机样本。先创建如下类变量：

    ```
    private static final int SAMPLE_COUNT = 50;
    private static final Random random = new Random();
    ```

3. 为 `SimpleBoxPlot` 类创建构造函数。

    ```
    public SimpleBoxPlot() {
    ```

4. 设置箱线图窗口尺寸。

    ```
    setPreferredSize(new Dimension(400, 600));
    ```

5. 创建一个数据表，每行包含 3 个列值，它们都是整型值。

    ```
    DataTable data = new DataTable(Integer.class, Integer.class,
        Integer.class);
    ```

6. 使用 3 个整型值（数据表中的列值）生成 50 个数据样本。数据样本包含的值服从高斯分布（这些数据不必非得服从高斯分布）。

    ```
    for (int i = 0; i < SAMPLE_COUNT; i++) {
    int x = (int) Math.round(5.0*random.nextGaussian());
    int y = (int) Math.round(5.0*random.nextGaussian());
    int z = (int) Math.round(5.0*random.nextGaussian());
    data.add(x, y, z);
    }
    ```

7. 使用数据新建一个箱线图。

    ```
    DataSource boxData = BoxPlot.createBoxData(data);
    BoxPlot plot = new BoxPlot(boxData);
    ```

8. 设置窗体的空白大小，我们将在该窗体中绘制箱线图。

    ```
    plot.setInsets(newInsets2D.Double(20.0, 50.0, 40.0, 20.0));
    ```

9. 为 x 轴上值设置格式。

    ```
    plot.getAxisRenderer(BoxPlot.AXIS_X).setCustomTicks(
    DataUtils.map(
    new Double[] {1.0, 2.0, 3.0},
    new String[] {"Column 1", "Column 2", "Column 3"}
    )
    );
    ```

10. 其他代码用来绘制箱线图。首先，使用数据创建一个 pointRenderer。

    ```
    BoxWhiskerRenderer pointRenderer =
        (BoxWhiskerRenderer) plot.getPointRenderer(boxData);
    ```

11. 接下来，设置箱线图的边框颜色、箱须颜色（第三个四分位数对应最大值，第一个四分位数对应最小值）与中间条（中位数线条）。

    ```
    pointRenderer.setBoxBorderColor(COLOR1);
    pointRenderer.setWhiskerColor(COLOR1);
    pointRenderer.setCenterBarColor(COLOR1);
    ```

12. 对箱线图使用垂直导航。

    ```
    plot.getNavigator().setDirection(XYNavigationDirection.VERTICAL);
    ```

13. 把箱线图添加到 Swing 组件，进行渲染。

    ```
    InteractivePanel panel = new InteractivePanel(plot);
    add(panel);
    ```

14. 关闭构造函数。

    ```
    }
    ```

15. 接下来，我们需要实现 ExamplePanel 类中的如下两个方法，把它们重载如下：

    ```
    @Override
    ```

```
    public String getTitle() {
    return "Box-and-whisker plot";
    }
    @Override
    public String getDescription() {
    return String.format("Three box-and-whisker plots created from
    %d random samples", SAMPLE_COUNT);
    }
```

16. 然后，添加 main 方法，关闭类。

```
    public static void main(String[] args) {
    new SimpleBoxPlot().showInFrame();
    }
    }
```

17. 完整的代码整理如下：

```
import java.awt.Dimension;
import java.util.Random;
import de.erichseifert.gral.data.DataSource;
import de.erichseifert.gral.data.DataTable;
import de.erichseifert.gral.examples.ExamplePanel;
import de.erichseifert.gral.plots.BoxPlot;
import de.erichseifert.gral.plots.BoxPlot.BoxWhiskerRenderer;
import de.erichseifert.gral.plots.XYPlot.XYNavigationDirection;
import de.erichseifert.gral.ui.InteractivePanel;
import de.erichseifert.gral.util.DataUtils;
import de.erichseifert.gral.util.Insets2D;

public class SimpleBoxPlot extends ExamplePanel {
    /** 序列化版本 id */
    private static final long serialVersionUID = 5228891435595348789L;
    private static final int SAMPLE_COUNT = 50;
    private static final Random random = new Random();

    @SuppressWarnings("unchecked")
    public SimpleBoxPlot() {
        setPreferredSize(new Dimension(400, 600));

        //创建样本数据
        DataTable data = new DataTable(Integer.class, Integer.class,
          Integer.class);
```

```java
    for (int i = 0; i < SAMPLE_COUNT; i++) {
        int x = (int) Math.round(5.0*random.nextGaussian());
        int y = (int) Math.round(5.0*random.nextGaussian());
        int z = (int) Math.round(5.0*random.nextGaussian());
        data.add(x, y, z);
    }

    //新建箱线图
    DataSource boxData = BoxPlot.createBoxData(data);
    BoxPlot plot = new BoxPlot(boxData);

    //设置图形格式
    plot.setInsets(new Insets2D.Double(20.0, 50.0, 40.0, 20.0));

    //设置坐标轴格式
    plot.getAxisRenderer(BoxPlot.AXIS_X).setCustomTicks(
        DataUtils.map(
            new Double[] {1.0, 2.0, 3.0},
            new String[] {"Column 1", "Column 2", "Column 3"}
        )
    );

    //设置箱盒
    /*Stroke stroke = new BasicStroke(2f);
    ScaledContinuousColorMapper colors =
        new LinearGradient(GraphicsUtils.deriveBrighter(COLOR1),
        Color.WHITE);
    colors.setRange(1.0, 3.0);*/

    BoxWhiskerRenderer pointRenderer =
            (BoxWhiskerRenderer) plot.getPointRenderer(boxData);
    /*pointRenderer.setWhiskerStroke(stroke);
    pointRenderer.setBoxBorderStroke(stroke);
    pointRenderer.setBoxBackground(colors);*/
    pointRenderer.setBoxBorderColor(COLOR1);
    pointRenderer.setWhiskerColor(COLOR1);
    pointRenderer.setCenterBarColor(COLOR1);

    plot.getNavigator().setDirection(XYNavigationDirection.VERTICAL);
    //把图形添加到Swing组件
    InteractivePanel panel = new InteractivePanel(plot);
    add(panel);
}
```

```
    @Override
    public String getTitle() {
        return "Box-and-whisker plot";
    }
    @Override
    public String getDescription() {
        return String.format("Three box-and-whisker plots created from %d
        random samples", SAMPLE_COUNT);
    }

    public static void main(String[] args) {
        new SimpleBoxPlot().showInFrame();
    }
}
```

9.6 绘制散点图

本节演示如何使用 GRAL 为 100 000 个随机数据点绘制散点图。散点图使用 *x* 轴与 *y* 轴来绘制数据点，这是一种用来展示两个变量之间相关性的好方法。

准备工作

1. 为了使用 GARL 绘制散点图，我们需要随该库一起提供的示例程序，这些示例以 Jar 文件形式存在。你可以从 `http://trac.erichseifert.de/gral/wiki/Download` 页面把 `gral-examples-0.10.zip` 下载到本地磁盘中，然后进行解压缩。

2. 下载好 Zip 文件之后，进行解压缩，在解压得到的目录结构（请参考 9.2 节中的准备部分）中，包含一个 `lib` 文件夹，该文件夹正是我们所感兴趣的。

3. 在 `lib` 文件夹中，包含 `gral-core-0.10`、`gralexamples-0.10`、`VectorGraphics2D-0.9.1` 3 个 Jar 文件。本节，我们将使用前两个 Jar 文件，即 `gral-core-0.10` 与 `gralexamples-0.10`。

4. 把 `gral-core-0.10` 与 `gralexamples-0.10` 两个 Jar 文件作为外部库添加到项目中。

接下来，我们就可以使用 GRAL 示例包中的程序来绘制散点图了。在下载示例包的 `gralexamples-0.10\gralexamples-0.10\src\main\java\de\erichseifert`

\gral\examples\xyplot 中，你可以看到我们要用的程序。成功运行本节代码后，你将会看到为 100 000 个随机数据点绘制的散点图，如图 9-13 所示。

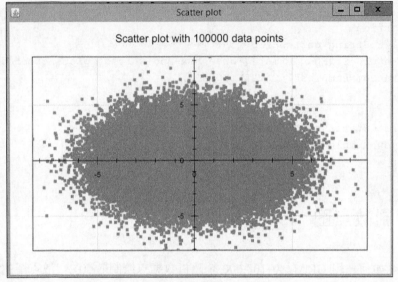

图 9-13

操作步骤

1. 首先，创建一个名为 ScatterPlot 的类，它继承了 GRAL 库中的 ExamplePanel 类。并且，在类中添加序列化版本 UID。

    ```
    public class ScatterPlot extends ExamplePanel {
    private static final long serialVersionUID =
        -4126994306259538870L;
    ```

2. 本节中，我们将会用到 100 000 个随机数据点。为数据点创建类变量，一个是样本数，另一个是随机数发生器。

    ```
    private static final int SAMPLE_COUNT = 100000;
    private static final Random random = new Random();
    ```

3. 接下来，开始编写构造函数。

    ```
    publicScatterPlot() {
    ```

4. 首先，创建一个数据表，用来存放随机生成的 x 与 y 值，它们对应于散点图中的点。

x 与 y 值都是 double 类型，并且服从高斯分布。

   ```
   DataTable data = new DataTable(Double.class, Double.class);
   for (int i = 0; i <= SAMPLE_COUNT; i++) {
   data.add(random.nextGaussian()*2.0,
    random.nextGaussian()*2.0);
   }
   ```

5. 我们可以把散点图看作 XYPlot，因此让我们先创建一个 XYPlot 对象。

   ```
   XYPlot plot =newXYPlot(data);
   ```

6. 为图形设置尺寸，以及获取图形描述。

   ```
   plot.setInsets(new Insets2D.Double(20.0, 40.0, 40.0, 40.0));
   plot.getTitle().setText(getDescription());
   ```

7. 为数据点设置格式，并设置颜色。

   ```
   plot.getPointRenderer(data).setColor(COLOR1);
   ```

8. 最后，把图形添加到 Java Swing 组件，关闭构造函数。

   ```
   add(new InteractivePanel(plot), BorderLayout.CENTER);
   }
   ```

9. 由于 ScatterPlot 类继承 ExamplePanel 类，所以我们还需要实现如下两个方法：

   ```
   @Override
   public String getTitle() {
   return "Scatter plot";
   }
   @Override
   public String getDescription() {
   return String.format("Scatter plot with %d data points",
   SAMPLE_COUNT);
   }
   ```

10. 最后，编写 main 方法，以便运行代码，关闭类。

    ```
    public static void main(String[] args) {
    new ScatterPlot().showInFrame();
    ```

            }
        }

示例的完整整理如下：

```java
import java.awt.BorderLayout;
import java.util.Random;
import de.erichseifert.gral.data.DataTable;
import de.erichseifert.gral.examples.ExamplePanel;
import de.erichseifert.gral.plots.XYPlot;
import de.erichseifert.gral.ui.InteractivePanel;
import de.erichseifert.gral.util.Insets2D;

public class ScatterPlot extends ExamplePanel {
    /** 序列化版本 id*/
    private static final long serialVersionUID = -4126994306259538887L;

    private static final int SAMPLE_COUNT = 100000;
    /** 用于产生随机数据值的实例*/
    private static final Random random = new Random();

    @SuppressWarnings("unchecked")
    public ScatterPlot() {
        //生成 100000 个数据点
        DataTable data = new DataTable(Double.class, Double.class);
        for (int i = 0; i <= SAMPLE_COUNT; i++) {
            data.add(random.nextGaussian()*2.0,
                random.nextGaussian()*2.0);
        }

        //新建 XYPlot 对象
        XYPlot plot = new XYPlot(data);

        //设置图形
        plot.setInsets(new Insets2D.Double(20.0, 40.0, 40.0, 40.0));
        plot.getTitle().setText(getDescription());

        //设置数据点
        plot.getPointRenderer(data).setColor(COLOR1);

        //把图形添加到 Swing 组件中
        add(new InteractivePanel(plot), BorderLayout.CENTER);
    }
```

```
    @Override
    public String getTitle() {
        return "Scatter plot";
    }

    @Override
    public String getDescription() {
        return String.format("Scatter plot with %d data points",
            SAMPLE_COUNT);
    }
    public static void main(String[] args) {
        new ScatterPlot().showInFrame();
    }
}
```

9.7 绘制甜圈图

甜圈图是饼图的一种，也是一种流行的数据可视化技术，用来描述数据之间的比例关系。本节，我们将学习使用 GRAL 这个 Java 库为 10 个随机变量绘制甜圈图。

准备工作

1. 为了使用 GARL 绘制甜圈图，我们需要随该库一起提供的示例程序，这些示例以 Jar 文件形式存在。你可以从 http://trac.erichseifert.de/gral/wiki/Download 页面把 `gral-examples-0.10.zip` 下载到本地磁盘中，然后进行解压缩。

2. 下载好 Zip 文件之后，进行解压缩，在解压得到的目录结构（请参考 9.2 节中的准备部分）中，包含一个 lib 文件夹，该文件夹正是我们所感兴趣的。

3. 在 `lib` 文件夹中，包含 `gral-core-0.10`、`gralexamples-0.10`、`VectorGraphics2D-0.9.1` 3 个 Jar 文件。本节，我们将使用前两个 Jar 文件，即 `gral-core-0.10` 与 `gralexamples-0.10`。

4. 把 `gral-core-0.10` 与 `gralexamples-0.10` 两个 Jar 文件作为外部库添加到项目中。

接下来，我们就可以使用 GRAL 示例包中的程序来绘制甜圈了。在下载示例包的 `gralexamples-0.10\gralexamples-0.10\src\main\java\de\erichseifert`

\gral\examples\pieplot 中，你可以看到我们要用的程序。成功运行本节代码后，你将会看到为 10 个随机数据值绘制的甜圈图，如图 9-14 所示。

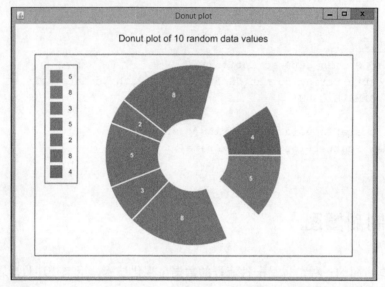

图 9-14

操作步骤

1. 创建一个名为 SimplePiePlot 的类，它继承了 GRAL 库中的 ExamplePanel 类。并在类中，声明一个序列化版本 UID 变量。

   ```
   public class SimplePiePlot extends ExamplePane {
   ```

2. 接着，声明两个类变量，用来生成 10 个随机数据点。

   ```
   privatestaticfinalintSAMPLE_COUNT = 10;
   privatestatic Random random = new Random();
   ```

3. 然后，开始为 SimplePiePlot 类编写构造函数。

   ```
   public SimplePiePlot() {
   ```

4. 创建一个数据表，并把 10 个随机数放入其中。本例中，我们先使用种子值 8 创建一个随机整数（由 Random 类生成），然后把该随机数与 2 相加得到 val 值。然后，检查随机生成的 double 值是否小于或等于 0.15。如果是，则把 value 的负值添加到数据表中，否则直接把 value 添加到数据表中。

```
DataTable data = new DataTable(Integer.class);
for (int i = 0; i < SAMPLE_COUNT; i++) {
int val = random.nextInt(8) + 2;
data.add((random.nextDouble() <= 0.15) ? -val : val);
}
```

5. 使用数据表（data）创建一个 `PiePlot`。

    ```
    PiePlot plot = new PiePlot(data);
    ```

6. 获取甜圈图的标题。

    ```
    plot.getTitle().setText(getDescription());
    ```

7. 接下来，设置甜圈图的相对尺寸。

    ```
    plot.setRadius(0.9);
    ```

8. 如果想在甜圈图中显示图例，只要将其可视性设置为 true 即可。否则，将其设置为 `false`。

    ```
    plot.setLegendVisible(true);
    ```

9. 为甜圈图设置尺寸。

    ```
    plot.setInsets(new Insets2D.Double(20.0, 40.0, 40.0, 40.0));
    ```

10. 为甜圈图创建一个 `PointRender`。

    ```
    PieSliceRenderer pointRenderer =
       (PieSliceRenderer) plot.getPointRenderer(data);
    ```

11. 设置内部区域的相对大小。

    ```
    pointRenderer.setInnerRadius(0.4);
    ```

12. 在两个切块之间设置合适的间隔。

    ```
    pointRenderer.setGap(0.2);
    ```

13. 设置切块颜色。

```
LinearGradient colors = new LinearGradient(COLOR1, COLOR2);
pointRenderer.setColor(colors);
```

14. 设置标签格式与显示方式。本例中，我们把显示的数字设置为白色粗体。

    ```
    pointRenderer.setValueVisible(true);
    pointRenderer.setValueColor(Color.WHITE);
    pointRenderer.setValueFont(Font.decode(null)
      .deriveFont(Font.BOLD));
    ```

15. 最后，把甜圈图添加到 Swing 组件中。

    ```
    add(new InteractivePanel(plot), BorderLayout.CENTER);
    ```

16. 关闭构造函数。

    ```
    }
    ```

17. 由于 `SimplePiePlot` 类继承了 `ExamplePanel` 类，因此我们还需要在代码中实现另外两个方法。

    ```
    @Override
    public String getTitle() {
    return "Donut plot";
    }
     @Override
      public String getDescription() {
      return String.format("Donut plot of %d random data values",
        SAMPLE_COUNT);
      }
    ```

18. 编写 main 方法，以便运行代码。

    ```
    publicstaticvoid main(String[] args) {
    new SimplePiePlot().showInFrame();
    }
    ```

19. 关闭类。

    ```
    }
    ```

20. 完整的示例代码整理如下：

```java
import java.awt.BorderLayout;
import java.awt.Color;
import java.awt.Font;
import java.util.Random;
import de.erichseifert.gral.data.DataTable;
import de.erichseifert.gral.examples.ExamplePanel;
import de.erichseifert.gral.plots.PiePlot;
import de.erichseifert.gral.plots.PiePlot.PieSliceRenderer;
import de.erichseifert.gral.plots.colors.LinearGradient;
import de.erichseifert.gral.ui.InteractivePanel;
import de.erichseifert.gral.util.Insets2D;

public class SimplePiePlot extends ExamplePanel {
    /** 序列化版本id*/
    private static final long serialVersionUID = -3039317265508932299L;

    private static final int SAMPLE_COUNT = 10;
    /** 用于产生随机数据值的实例*/
    private static Random random = new Random();

    @SuppressWarnings("unchecked")
    public SimplePiePlot() {

        //生成数据
        DataTable data = new DataTable(Integer.class);
        for (int i = 0; i < SAMPLE_COUNT; i++) {
            int val = random.nextInt(8) + 2;
            data.add((random.nextDouble() <= 0.15) ? -val : val);
        }

        //新建饼图
        PiePlot plot = new PiePlot(data);

        //设置图形
        plot.getTitle().setText(getDescription());
        //设置饼块的相对大小
        plot.setRadius(0.9);
        //显示图例
        plot.setLegendVisible(true);
        //向图形区域添加边距
        plot.setInsets(new Insets2D.Double(20.0, 40.0, 40.0, 40.0));

        PieSliceRenderer pointRenderer =
```

```
                (PieSliceRenderer) plot.getPointRenderer(data);
        //设置内区域相对大小
        pointRenderer.setInnerRadius(0.4);
        //设置饼块间的间隔大小
        pointRenderer.setGap(0.2);
        //设置颜色
        LinearGradient colors = new LinearGradient(COLOR1, COLOR2);
        pointRenderer.setColor(colors);
        //显示标签
        pointRenderer.setValueVisible(true);
        pointRenderer.setValueColor(Color.WHITE);
        pointRenderer.setValueFont(Font.decode(null).deriveFont(Font.BOLD));

        //把图形添加到Swing组件中
        add(new InteractivePanel(plot), BorderLayout.CENTER);
    }
    @Override
    public String getTitle() {
        return "Donut plot";
    }

    @Override
    public String getDescription() {
        return String.format("Donut plot of %d random data values",
            SAMPLE_COUNT);
    }
    public static void main(String[] args) {
        new SimplePiePlot().showInFrame();
    }
}
```

9.8 绘制面积图

在描述数量值一个给定区间上的发展趋势时,面积图是一种非常有用的工具。对数据科学家来说,面积图是用来了解某个发展趋势的有效工具。面积图基于折线图,但在线条(它们基于坐标轴上的值绘制而成)之下的区域中填充有某种颜色或纹理。本节,我们将学习使用GRAL Java库来绘制面积图。

准备工作

1. 为了使用 GARL 绘制面积图，我们需要随该库一起提供的示例程序，这些示例以 Jar 文件形式存在。你可以从 http://trac.erichseifert.de/gral/wiki/ Download 页面把 `gral-examples-0.10.zip` 下载到本地磁盘中，然后进行解压缩。

2. 下载好 Zip 文件之后，进行解压缩，在解压得到的目录结构（请参考 9.2 节中的准备部分）中，包含一个 `lib` 文件夹，该文件夹正是我们所感兴趣的。

3. 在 `lib` 文件夹中，包含 `gral-core-0.10`、`gralexamples-0.10`、`VectorGraphics2D-0.9.1` 3 个 Jar 文件。本节，我们将使用前两个 Jar 文件，即 `gral-core-0.10` 与 `gralexamples-0.10`。

4. 把 `gral-core-0.10` 与 `gralexamples-0.10` 两个 Jar 文件作为外部库添加到项目中。

接下来，我们就可以使用 GRAL 示例包中的程序来绘制面积图了。成功运行本节代码后，你将会看到如图 9-15 所示的面积图。

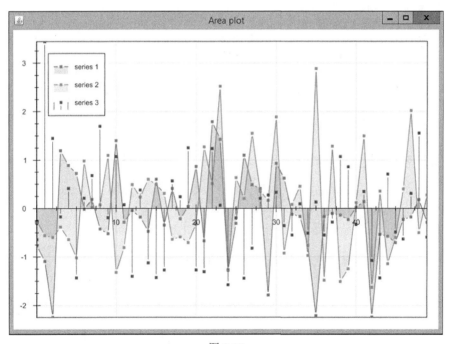

图 9-15

操作步骤

1. 首先，创建一个名为 `AreaPlot` 的类，它继承了 GRAL 的 `ExamplePanel` 类。并在类中，声明一个类序列化版本 UID。

   ```
   public class AreaPlot extends ExamplePanel {
   private static final long serialVersionUID =
     3287044991898775949L;
   ```

2. 我们将使用随机数来绘制面积图。因此，先创建一个用于生成随机数的类变量。

   ```
   private static final Random random = new Random();
   ```

3. 接下来，为类编写构造函数。

   ```
   public AreaPlot() {
   ```

4. 创建一个数据表，用于存放 4 个数据点：一个 *x* 值与 3 个 *y* 值。本例中数据点的所有值都是 Double 类型。

   ```
   DataTable data = new DataTable(Double.class, Double.class,
       Double.class, Double.class);
   ```

5. 创建一个循环 50 次的 for 循环，*x* 的初始值为 `0.0`，步长为 1。

   ```
   for (double x = 0.0; x < 50; x ++) {
   ```

6. 创建 3 个变量，用来存放 *y* 值。这些 *y* 值是随机生成的，并且服从高斯分布。

   ```
   y1 = random.nextGaussian();
   y2 = random.nextGaussian();
   y3 = random.nextGaussian();
   ```

7. 最后，把(x, y1, y2, y3)添加到数据表，关闭 for 循环。

   ```
   data.add(x, y1, y2, y3);
   }
   ```

为了得到更好的图形，我们可以使用如下代码代替步骤 5~7 中的 for 循环。

```
for (double x=0.0; x<.5*Math.PI; x+=Math.PI/15.0) {
 double y1 = Double.NaN, y2 = Double.NaN, y3 = Double.NaN;
 if (x>=0.00*Math.PI && x<2.25*Math.PI) {
    y1 = 4.0*Math.sin(x + 0.5*Math.PI) +
```

```
             0.1*random.nextGaussian();
    }
    if (x>=0.25*Math.PI && x<2.50*Math.PI) {
       y2 = 4.0*Math.cos(x + 0.5*Math.PI) +
             0.1*random.nextGaussian();
    }
    if (x>=0.00*Math.PI && x<2.50*Math.PI) {
       y3 = 2.0*Math.sin(2.0*x/2.5)
             0.1*random.nextGaussian();
    }
       data.add(x, y1, y2, y3);
    }
```

8. 然后，使用 GRAL 的 `DataSeries` 类添加 3 组数据系列。`DataSeries` 类的构造函数如下：

   ```
   public DataSeries(DataSource data, int... cols)
   ```

9. 查阅 GRAL 的 Java API 文档可知，第一列是列 0，第二列是列 1，以此类推，而指定的列值则是数据源中的列号。

   ```
   DataSeries data1 = new DataSeries("series 1", data, 0, 1);
   DataSeries data2 = new DataSeries("series 2", data, 0, 2);
   DataSeries data3 = new DataSeries("series 3", data, 0, 3);
   ```

10. 使用 3 个数据系列，创建一个 `XYPlot`，在图形中把图例显示出来，并且设置图形的尺寸。

    ```
    XYPlot plot = new XYPlot(data1, data2, data3);
    plot.setLegendVisible(true);
    plot.setInsets(new Insets2D.Double(20.0, 40.0, 20.0, 20.0));
    ```

11. 绘制面积图时，还需要进行的一道"工序"是用颜色进行填充。我们将调用名为 `formatFilledArea` 与 `formatLineArea` 的静态方法来完成这项任务。请注意前两个数据系列与第三个系列的不同之处。

    ```
    formatFilledArea(plot, data1, COLOR2);
    formatFilledArea(plot, data2, COLOR1);
    formatLineArea(plot, data3, GraphicsUtils.deriveDarker(COLOR1));
    ```

12. 把图形添加到 Swing 组件，然后关闭构造函数。

```
        add(new InteractivePanel(plot));
    }
```

13. 创建一个静态方法,用来使用某种颜色填充区域。这个方法有 3 个参数,分别是 XYPlot、数据系列、颜色。

```
    private static void formatFilledArea(XYPlot plot, DataSource
        data, Color color) {
```

14. 创建一个 PointRenderer。请注意,由于我们要绘制的是 2D 图形,所以要选择合适的类。先为 PointRenderer 设置颜色,而后为数据系列设置 PointRenderer。

```
        PointRenderer point = new DefaultPointRenderer2D();
        point.setColor(color);
        plot.setPointRenderer(data, point);
```

15. 类似地,使用 GRAL 中相应类创建一个 2D LineRenderer,为其设置颜色,把线条间隔设置为 3.0。接着,把 GapRounded 设置为真。最后,把 LineRenderer 设置给数据系列。

```
        LineRenderer line = new DefaultLineRenderer2D();
        line.setColor(color);
        line.setGap(3.0);
        line.setGapRounded(true);
        plot.setLineRenderer(data, line);
```

16. 接下来,我们还要绘制区域。创建一个 2D AreaRenderer,为它设置颜色。然后把 AreaRenderer 设置给数据系列,关闭方法。

```
        AreaRenderer area = new DefaultAreaRenderer2D();
        area.setColor(GraphicsUtils.deriveWithAlpha(color, 64));
        plot.setAreaRenderer(data, area);
    }
```

17. 类似地,创建一个静态方法对折线区域进行设置。该方法有 3 个参数,分别是 XYPlot(已在构造函数中创建好)、数据系列、颜色。

```
    private static void formatLineArea(XYPlot plot, DataSource
        data, Color color) {
```

18. 创建一个 2D PointRenderer,为其设置颜色,并把它设置给数据系列。

```
PointRenderer point = new DefaultPointRenderer2D();
point.setColor(color);
plot.setPointRenderer(data, point);
```

19. 在这个方法中,我们不会使用 LineRenderer。这将会让第三个数据系列看上去与前两个数据系列不同。

    ```
    plot.setLineRenderer(data, null);
    ```

20. 参考前面步骤,创建一个 2D AreaRenderer,设置区域间隔、颜色,把这个 AreaRenderer 设置给数据系列。

    ```
    AreaRenderer area = new LineAreaRenderer2D();
    area.setGap(3.0);
    area.setColor(color);
    plot.setAreaRenderer(data, area);
    }
    ```

21. 接下来,我们需要重载 ExamplePanel 类的两个方法,如下:

    ```
    @Override
    public String getTitle() {
    return "Area plot";
    }
    @Override
    public String getDescription() {
    return "Area plot of three series with different styling";
    }
    ```

22. 为了运行上面代码,需要编写如下 main 方法。最后,关闭类。

    ```
    public static void main(String[] args) {
    new AreaPlot().showInFrame();
    }
    }
    ```

示例的完整代码整理如下:

```
import java.awt.Color;
import java.util.Random;
import de.erichseifert.gral.data.DataSeries;
import de.erichseifert.gral.data.DataSource;
```

```java
import de.erichseifert.gral.data.DataTable;
import de.erichseifert.gral.examples.ExamplePanel;
import de.erichseifert.gral.plots.XYPlot;
import de.erichseifert.gral.plots.areas.AreaRenderer;
import de.erichseifert.gral.plots.areas.DefaultAreaRenderer2D;
import de.erichseifert.gral.plots.areas.LineAreaRenderer2D;
import de.erichseifert.gral.plots.lines.DefaultLineRenderer2D;
import de.erichseifert.gral.plots.lines.LineRenderer;
import de.erichseifert.gral.plots.points.DefaultPointRenderer2D;
import de.erichseifert.gral.plots.points.PointRenderer;
import de.erichseifert.gral.ui.InteractivePanel;
import de.erichseifert.gral.util.GraphicsUtils;
import de.erichseifert.gral.util.Insets2D;

public class AreaPlot extends ExamplePanel {
    /** 序列化版本 id*/
    private static final long serialVersionUID = 3287044991898775949L;

    /** 用于产生随机数据值的实例*/
    private static final Random random = new Random();

    public AreaPlot() {
        //生成数据
        DataTable data = new DataTable(Double.class, Double.class,
            Double.class, Double.class);
        for (double x = 0.0; x < 50; x ++) {
            double y1 = Double.NaN, y2 = Double.NaN, y3 = Double.NaN;
            y1 = random.nextGaussian();
            y2 = random.nextGaussian();
            y3 = random.nextGaussian();
            data.add(x, y1, y2, y3);
        }

        //创建数据系列
        DataSeries data1 = new DataSeries("series 1", data, 0, 1);
        DataSeries data2 = new DataSeries("series 2", data, 0, 2);
        DataSeries data3 = new DataSeries("series 3", data, 0, 3);

        //新建 XYPlot 对象
        XYPlot plot = new XYPlot(data1, data2, data3);
        plot.setLegendVisible(true);
        plot.setInsets(new Insets2D.Double(20.0, 40.0, 20.0, 20.0));
```

```java
        //设置数据系列
        formatFilledArea(plot, data1, COLOR2);
        formatFilledArea(plot, data2, COLOR1);
        formatLineArea(plot, data3, GraphicsUtils.deriveDarker(COLOR1));

        //把图形添加到Swing组件中
        add(new InteractivePanel(plot));
    }

    private static void formatFilledArea(XYPlot plot, DataSource data,
        Color color) {
        PointRenderer point = new DefaultPointRenderer2D();
        point.setColor(color);
        plot.setPointRenderer(data, point);
        LineRenderer line = new DefaultLineRenderer2D();
        line.setColor(color);
        line.setGap(3.0);
        line.setGapRounded(true);
        plot.setLineRenderer(data, line);
        AreaRenderer area = new DefaultAreaRenderer2D();
        area.setColor(GraphicsUtils.deriveWithAlpha(color, 64));
        plot.setAreaRenderer(data, area);
    }
    private static void formatLineArea(XYPlot plot, DataSource data,
        Color color) {
        PointRenderer point = new DefaultPointRenderer2D();
        point.setColor(color);
        plot.setPointRenderer(data, point);
        plot.setLineRenderer(data, null);
        AreaRenderer area = new LineAreaRenderer2D();
        area.setGap(3.0);
        area.setColor(color);
        plot.setAreaRenderer(data, area);
    }

    @Override
    public String getTitle() {
        return "Area plot";
    }

    @Override
    public String getDescription() {
        return "Area plot of three series with different styling";
```

```
    }

    public static void main(String[] args) {
        new AreaPlot().showInFrame();
    }
}
```